Hansjörg Seybold

Modeling River Delta Formation

Hansjörg Seybold

Modeling River Delta Formation

Computations, Experiments and Observations

Südwestdeutscher Verlag für Hochschulschriften

Impressum/Imprint (nur für Deutschland/ only for Germany)
Bibliografische Information der Deutschen Nationalbibliothek: Die Deutsche Nationalbibliothek verzeichnet diese Publikation in der Deutschen Nationalbibliografie; detaillierte bibliografische Daten sind im Internet über http://dnb.d-nb.de abrufbar.

Alle in diesem Buch genannten Marken und Produktnamen unterliegen warenzeichen-, marken- oder patentrechtlichem Schutz bzw. sind Warenzeichen oder eingetragene Warenzeichen der jeweiligen Inhaber. Die Wiedergabe von Marken, Produktnamen, Gebrauchsnamen, Handelsnamen, Warenbezeichnungen u.s.w. in diesem Werk berechtigt auch ohne besondere Kennzeichnung nicht zu der Annahme, dass solche Namen im Sinne der Warenzeichen- und Markenschutzgesetzgebung als frei zu betrachten wären und daher von jedermann benutzt werden dürften.

Verlag: Südwestdeutscher Verlag für Hochschulschriften Aktiengesellschaft & Co. KG
Dudweiler Landstr. 99, 66123 Saarbrücken, Deutschland
Telefon +49 681 37 20 271-1, Telefax +49 681 37 20 271-0
Email: info@svh-verlag.de
Zugl.: Zurich, ETH, Diss. 2009

Herstellung in Deutschland:
Schaltungsdienst Lange o.H.G., Berlin
Books on Demand GmbH, Norderstedt
Reha GmbH, Saarbrücken
Amazon Distribution GmbH, Leipzig
ISBN: 978-3-8381-1466-8

Imprint (only for USA, GB)
Bibliographic information published by the Deutsche Nationalbibliothek: The Deutsche Nationalbibliothek lists this publication in the Deutsche Nationalbibliografie; detailed bibliographic data are available in the Internet at http://dnb.d-nb.de.

Any brand names and product names mentioned in this book are subject to trademark, brand or patent protection and are trademarks or registered trademarks of their respective holders. The use of brand names, product names, common names, trade names, product descriptions etc. even without a particular marking in this works is in no way to be construed to mean that such names may be regarded as unrestricted in respect of trademark and brand protection legislation and could thus be used by anyone.

Publisher: Südwestdeutscher Verlag für Hochschulschriften Aktiengesellschaft & Co. KG
Dudweiler Landstr. 99, 66123 Saarbrücken, Germany
Phone +49 681 37 20 271-1, Fax +49 681 37 20 271-0
Email: info@svh-verlag.de

Printed in the U.S.A.
Printed in the U.K. by (see last page)
ISBN: 978-3-8381-1466-8

Copyright © 2010 by the author and Südwestdeutscher Verlag für Hochschulschriften Aktiengesellschaft & Co. KG and licensors
All rights reserved. Saarbrücken 2010

Diss. ETH No. 18263

Modeling River Delta Formation

A dissertation submitted to the
SWISS FEDERAL INSTITUTE OF TECHNOLOGY (ETH)
ZÜRICH

for the degree of
DOCTOR OF SCIENCE

presented by
HANSJÖRG FLORIAN SEYBOLD

born 28.10.1976
citizen of Germany

accepted on the recommendation of
Prof. Dr. H. J. Herrmann, examiner
Prof. Dr. W. Kinzelbach, co-examiner
Prof. Dr. H. Seyfried, co-examiner

August 1, 2009

Title: Satellite image of the Lena Delta, Russia. Landsat, Earth as Art series.
©USGS.gov/NASA.

Contents

1	**Introduction**	**1**
2	**River Deltas**	**7**
2.1	Dynamics of River Delta Growth	7
2.2	Classification of River Deltas	12
	2.2.1 River-dominated Deltas	13
	2.2.2 Wave-dominated Deltas	17
	2.2.3 Tide-dominated Deltas	20
3	**Coastal and River Modeling**	**23**
3.1	Hydrodynamic Models	24
3.2	Reduced Complexity Models	27
4	**A Delta Formation Model**	**31**
4.1	The Model	32
4.2	Simulation	36
4.3	Modeling different delta types	39
4.4	Interpretation of the model parameters	47
	4.4.1 Water-sediment fluxes and lobe volume	49
	4.4.2 Comparison of variables and parameters with the Balize Lobe	52
	4.4.3 The erosion-sedimentation process	53
4.5	Lobe Switching and Long-term temporal Dynamics	59
5	**Inland Delta generation**	**67**
5.1	The Okavango Delta	67
	5.1.1 Hydrology of the Okavango	68
	5.1.2 Geology of the Okavango	70
	5.1.3 Sedimentation processes in the Okavango Delta	71
5.2	Modeling Inland Delta Formation	73

	5.2.1	Modification of the River Delta Model	73
	5.2.2	Simulating Inland Delta Formation	74
	5.2.3	Rescaling of the variables	79
5.3	Experimental Modeling		81

6 Conclusions 95

List of Figures

1.1	The Nile Delta	2
1.2	Important river deltas in the world	2
2.1	Components of the delta plain	8
2.2	Cross-section through the delta	8
2.3	Deposition at the delta mouth	9
2.4	Galloway classification triangle	13
2.5	River dominated deltas	13
2.6	Drainage area of the Mississippi River	14
2.7	Main lobes of the Mississippi River Delta	16
2.8	Satellite image of the currently active lobe of the Mississippi Delta	17
2.9	Wave dominated deltas	17
2.10	São Francisco Delta	19
2.11	Tide dominated deltas	20
2.12	Ganges-Brahmaputra Delta	21
2.13	Fly Delta	22
4.1	Model river segment	33
4.2	Initial and boundary conditions of the simulation	37
4.3	Schematic description of the model cycle	38
4.4	Comparison of the simulation with the Mississippi Delta	40
4.5	Comparison of the simulation with the São Francisco Delta	41
4.6	Comparison of the simulation with lobate shaped deltas	41
4.7	Fractal dimension of the simulation and the Lena Delta	42
4.8	Time evolution of the fractal dimension during the simulation	43
4.9	Fractal dimension of wave dominted deltas	44
4.10	Fractal dimension of birdfoot deltas	44
4.11	Growth of the fractal dimension of a birdfoot delta simulation	45
4.12	Size distribution of island in the simulated and the Lena Delta	46

4.13	Size distribution of islands in the simulation of a birdfoot delta and the Mississippi	47
4.14	Histogram of the island eccentricity	48
4.15	Reconstruction of the ancient Mississippi Delta mouth	51
4.16	Levee formation in the simulation of birdfoot deltas	54
4.17	Cross-section through the Mississippi Balize lobe	54
4.18	Subaqueous and subaerial levee formation in the simulation	55
4.19	Histogram of water and sediment flow	56
4.20	Frequency of the sedimentation-/erosion rates	57
4.21	Lobe switching	59
4.22	Erosion & sediment flux at the delta mouth	60
4.23	Time evolution of a birdfoot delta	62
4.24	DFA of delta growth	65
5.1	Map of the Okavango Delta	68
5.2	Aerial photo of the Okavango Delta	70
5.3	Sedimentation in channel ends	74
5.4	Parameter study of inland delta formation	76
5.5	Cumulative channel width	77
5.6	Channel cross-section	77
5.7	Sedimentation in inland deltas	78
5.8	Fractal dimension of the simulated delta and the Okavango	78
5.9	Experimental setup	82
5.10	Experimental setup and 3D scanning	82
5.11	Experimental deposition pattern	83
5.12	Levee formation in the experiment	84
5.13	Elevation statistics	85
5.14	Channel cross-section of the simulation	86
5.15	Deposition pattern of the experiment	86
5.16	Deposition pattern of the experiment	87
5.17	Slope & aspect	88
5.18	Surface change of the experiment	89
5.19	Slope statistics of the experiment	90
5.20	Slope statistics of the Okavango case	91
5.21	Slope statistics of the inland delta simulation	92
5.22	Slope statistics of the Mississippi Balize Lobe	92
5.23	Slope statistics of the birdfoot delta simulation	93

Zusammenfassung

Wie entstehen Flussdeltas und wie verändern sie sich mit der Zeit? Diese klassischen Fragen der Geologie haben sich in der physikalischen Modellbildung als ein sehr hartnäckiges Problem herausgestellt. Während hydrodynamische Gleichungen den Verlauf des Wasserflusses über eine sehr kurze Zeit hinweg recht genau beschreiben, ist die Modellierung über geologische Zeiträume hinweg wegen des immens hohen Rechenaufwands bis heute nicht möglich. Daher ist es notwendig, die Komplexität der Gleichungen auf die essentiellen Prozesse zu reduzieren, ohne jedoch die Charakteristik des Systems zu verändern.
In den letzten Jahren haben sich sogenannte Zellularautomaten als eine erfolgreiche Methode in der geomorphologischen Modellierung herausgestellt. Während diese Modelle jedoch z.B. auf verzopfte Flüsse sehr erfolgreich angewandt werden konnten, war es bisher nicht möglich, realistische Flussdeltas zu simulieren. Dies liegt vor allem daran, dass die verwendeten Flussgleichungen wie Manning-Strickler etc. nicht auf ozeanische Bewegungen angewandt werden können.
Basierend auf dem Prinzip von Zellularautomaten beschreibt das in diesem Projekt vorgestellte Modell die Entwicklung eines Flussdeltas über einen geologischen Zeitraum hinweg und erlaubt so mit Hilfe des Computers einen tieferen Einblick in die Prozesse, die an der Deltaentwicklung beteiligt sind. Dieses neue Modell kombiniert die Einfachheit der zellulären Modelle mit einer (vereinfachten) hydrodynamischen Modellierung, die für eine realistische Beschreibung der Deltaentwicklung notwendig ist.
Die Topographie wird dabei auf einem Quadratgitter diskretisiert, wobei auf jedem Knoten die Höhe der Landschaft und der Wasserstand, bezogen auf Meeresniveau, gespeichert sind. Diese Knoten werden durch Kanten mit einer hydrologischen Leitfähigkeit verbunden, die bestimmt, wieviel Wasser von einem Knoten zum anderen fliessen kann. Dies wiederum hängt von der Wassertiefe an den beteiligten Knoten ab. Der Wasserfluss von einem Knoten zum anderen ergibt sich nun als Produkt der Leitfähigkeit

mit dem Gefälle des Wasserpegels an den benachbarten Knoten. Ist der Wasserpegel hoch oder das Gefälle gross, so kann mehr Flüssigkeit transportiert werden als in einem flachen Gewässer. Des weiteren hängt der Durchfluss, der von einem zum anderen Knoten fliesst, vom hydrostatischen Druck zwischen den Endknoten ab, der durch das Gefälle des Wasserspiegels in den beteiligten Knoten beschrieben wird. Der gesamte Fluss ergibt sich somit als das Produkt aus Leitfähigkeit und Gefälle. Die Beschreibung des Wasserflusses wird durch die Massenerhaltung in jedem Knoten vervollständigt. In dieser Arbeit wurde das Modell auf dem Computer implementiert und getestet, sowie die Abhängigkeit der Kanalmuster von den einzelnen Parametern des Modells untersucht. Je nach Stärke des Wasserflusses auf einer Kante wird entweder erodiert oder sedimentiert. Dabei gilt, je stärker der Wasserfluss auf einer Kante ist, umso mehr Material kann abgetragen werden. Erodiertes Material wird mit dem Wasserstrom fort transportiert. Während sich auf den inneren Knoten sowohl die Landschaft und der Wasserspiegel ändern können, wird am Rand an der Meerseite der Wasserspiegel konstant auf null gehalten. Dadurch kann Wasser aus dem System abfliessen. Auf der anderen Seite wird ein konstanter Wasser- und Sedimentstrom in das System eingespeist. Dabei konnten interessante Übereinstimmungen mit echten Deltas gefunden werden, sowohl in den einzelnen Deltaformen, als auch in der Zeitentwicklung.

Diese Resultate wurden bereits in einer Publikation in "Proceedings of the National Academy of Science, USA", sowie im "Journal of Geophysical Research" publiziert. Eine weitere Publikation wurde in der Zeitschrift "Int. J. Mod. Phys.B" veröffentlicht. Grundsätzlich unterscheidet man in der Geologie drei wesentliche Faktoren, die über die Form des Deltas entscheiden. Neben dem Fluss selbst sind dies die Gezeiten und die Kraft der Wellen des Gewässers, in das der Fluss mündet. Durch Modifikation des Erosions-/Sedimentationsgesetzes konnten die verschiedenen Deltatypen in der Simulation reproduziert werden, wobei die Ähnlichkeit anhand der fraktalen Dimension des Kanalmusters quantifiziert wurde. Des weiteren konnte in der Zeitentwicklung das bei natürlichen Deltas typische Phänomen des "Lobe switchings" beobachtet werden. Dies tritt auf, wenn der Fluss durch anhaltende Sedimentation an der Deltamündung an seinem Abfluss blockiert wird und sich daraufhin einen neuen Verlauf suchen muss. Dieses Phänomen ist eine intrinsische Eigenschaft des Modells und wird nicht von aussen durch einen Switching-Parameter induziert, wie es bei anderen Modellen der Fall ist. Die Schlüssigkeit des Modells konnte am Beispiel des Balize Lobes (Mississippi Delta) verifiziert werden indem Zeit- und Längenskalen der dimensionslosen Modellgleichungen reskaliert und mit gemessenen Grössen verglichen wurden.

Im Rahmen des SNF Projekts 200021-116050 wurde das Modell auf das Binnendelta des Okavango angewandt. Dazu wurde das Modell um Evaporation und Versickerung erweitert. Die entstehenden Ablagerungen zeigen grosse Ähnlichkeiten mit den Ablagerungen im Okavango und auch die Ausbildung von natürlichen "Levees" konnte im Modell reproduziert werden. Die Simulationen wurden durch ein Flume Experiment ergänzt, das die Bedingungen im Okavango mit Evaporation auf Laborskala nachstellen soll. Die Topographie wurde zu verschiedenen Zeitpunkten mit einem 3D Scanner vermessen, so dass später eine dreidimensionale Rekonstruktion des Sedimentationsmusters möglich ist. Ziel war es durch Vergleich von Simulation, Experiment und Feldmessungen, die geomorphologischen Unterschiede von Trockendeltas und Küstendeltas zu charakterisieren, um so einen tieferen Einblick in die unterschiedlichen Sedimentationsprozesse in den beiden Deltaformen und im speziellen im Okavango zu erhalten.

1. August, 2009
Hansjörg Seybold

Summary

How do river deltas emerge and how do they evolve with time? These classical questions in geology are to be computationally difficult to answer due to the long times and large ranges of length scales involved in the processes. While classical hydrodynamic models based on physical equations describe water flow in channels quite well they are not capable to capture processes taking place over geological time scales (many thousands or even millions of years). Therefore, it is not only necessary to reduce complexity but also to include controls for all factors dominating changes in morphology.

In recent years, cellular automata models have been adopted in geomorphological modeling. While they have been successfully applied to braided river systems and other topography-driven flows,they failed to reproduce the characteristic features of coastal deltas. This comes mainly from the fact, that coastal currents cannot be described by topographic flow routing rules like the Manning-Strickler formula.

Based on the ideas of cellular automata, a new model simulating river delta evolution is proposed in this study, providing insight into the theoretical processes of the delta formation. The model combines the simplicity of cellular models with the essential hydrodynamic features necessary to reproduce realistic river delta patterns.

The model discretizes the topography on a square lattice where the elevation of the surface and water level are defined on the nodes. The nodes are connected with their nearest neighbors through hydraulic conductivity bonds determined by the water depth on corresponding end nodes. The amount of flow from one node to the other is determined by the product of hydraulic conductivity and difference of water levels in adjacent nodes. If the water is deep or the gradient steep, more water can be transported through the bond than in a shallow channel. In each node water mass is conserved using the continuity equation.

In this study, a computational smulation of the model is presented. The parameters have been adjusted the resulting channel pattern are compared with natural delta features. Water and sediment are injected into an entrance node. Depending on the strength of the water flow the surface is eroded or sediment is deposited on the channel bottom; eroded sediment is advected by the water flow. While the water level on the inner nodes is variable the sea level on the far ocean boundary is kept fixed, thus water and sediment can leave the domain through the ocean edges. Simulated ocean deltas show comparable features to real deltas with respect to surface pattern and major evolutionary changes. Furthermore the subaqueous morphology and the formation of levees could be reproduced for the first time with a reduced complexity model.

Three main types of deltas are distinguished in geology according to the dominnant forcess that shape them, namely river, wave and tide dominated deltas. These delta types can be reproduced in the simulation through modification of the sedimentation- and erosion laws. The similarity to real deltas is quantified by the fractal dimension of the delta pattern.

Dynamics of the deep-time evolution of real deltas, such as lobe switching, are reproduced in the model too. The switching of the locus of the delta lobe occurs when the sediment blocks the flow of the river on its way to the ocean and the river has to find a new course. This phenomenon is an intrinsic property of the model dynamics and is not introduced by a switching parameter as in other models. By rescaling length and time scales and comparing the properties with the evolution of the Balize Lobe of the Mississippi we show that the model equations are consistent and that predicted deposition volumes agree with what is observed in nature.

In the SNF Project 200021-116050, the model equations have been extended to simulate the evolution of the Okavango Delta in southwestern Africa. Simulations of the special conditions governing this type of inland delta requires the inclusion of additional controlling factors into the model such as evaporation and seepage. The obtained sedimentation patterns show close similarity to the patterns observed in the Okavango Delta.

The model reproduces details of sedimentary dynamics such as the formation of natural levees. The simulations have been accompanied by a flume experiment that reproduces conditions in the Okavango, including evaporation on a laboratory scale. At different times, the topography was measured using a sterioscopic scanning device. With the three dimensional information of the topography over time, it is possible to reconstruct the different facies of the sedimentation pattern.

By comparing simulation, experiment and field measurements, it is possible to gain a deeper insight into the different sedimentation processes in coastal

and inland deltas and especially in the Okavango Delta.
The results have been published in the "Proceedings of the National Academy of Science, USA", the "Journal of Geophysical Research", "Int. J. Mod .Phys.B", and "Geophysical Research Letters".

August 1, 2009
Hansjörg Seybold

Chapter 1

Introduction

Landforms and fluvial basins are the product of millions of years of tectonic movements coupled with erosion and weathering, which in turn strongly depend on the climate. River mouths are sites of constantly changing sedimentary accretion or erosion. In approximately 450 B.C. the great historian Herodotus used the term *delta* to describe the Nile River Delta because the sediment deposit at its mouth had the shape of upper-case Greek letter Delta: Δ. A satelite picture of the mouth of the Nile River is shown in Fig.1.1. Formally a delta can be defined as a coastal sedimentary deposit with both subaerial and subaqueous parts. It is formed by a river depositing its sediment at the edge of a standing body of water, in most cases an ocean also a lake. Sediment builds up as deltaic deposits when the river encounters a flat area that slows down te current.

Deltaic deposits of larger, heavily-laden rivers are characterized by a main channel dividing into multiple streams, known as distributaries, which divide and rejoin to form a maze of active and inactive channels. Deltas are also controlled by the width and slope of the shelf and sea level changes making the entire system extremely mobile. Twenty one of the world's twenty five largest rivers, which deliver 31% of total fluvial sediment reaching the ocean, have well-developed deltas at the open coast [70]. A map with the most important river deltas in the world is shown in Fig.1.2.

Deltas have high natural agricultural productivity, rich biodiversity and an abundance of waterways providing an easy means of transportation. As a result, 25% of the world's population live on deltaic coastlines and wetlands [70]. Deltas act as filters, repositories, and reactors for a host of continental materials such as mineral sediments, organic carbon, nutrients and pollutants, significantly affecting both the regional environment at the continent-ocean boundary, as well as global biochemical cycles. At

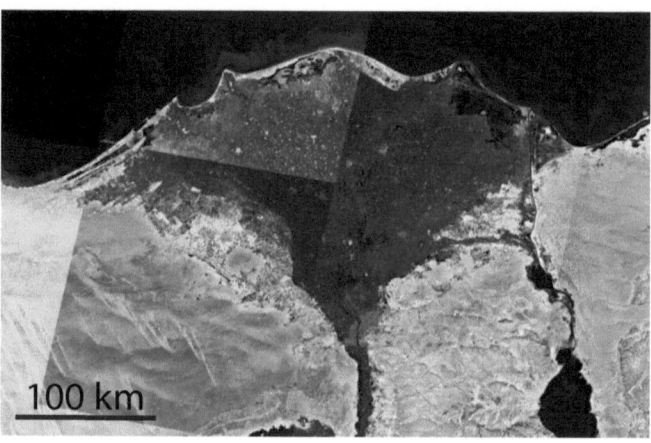

Fig. 1.1: *Satellite Picture of the Nile delta with the typical triangular shape. The picture was produced using NASA Worldwind™*

Fig. 1.2: *Important river deltas in the world. The largest deltas in the world are the Brahmaputra-Ganges delta in Bangla-Desh, the delta of the Amazon River, the Mississippi Delta and the delta of the Nile river. The classical end members of the classification scheme of Galloway are also shown in the map: the Mississippi as the most river-dominated delta, the Fly River in New Guinea as the delta strongest influenced by tidal currents and the São Francisco Delta in Brasil which is the most wave-dominated delta. One of the largest inland deltas in the world is the Okavango Delta in Botswana. The location of the Lena Delta shown on the title page is also marked on the map.*

the same time, deltas are fragile geomorphic structures, where modest modifications of the boundary conditions can have dramatic effects. Their complex structure is the result of the interplay between climatic-driven fluvial discharge, marine energy conditions, subsidence or uplift, and sea-level changes ranging from centimeters on the order of 10^3 years to -$120m/+270m$ on timescales of 10^4-$10^8 a$. All factors are in dynamic disequilibrium resulting in constant autocyclic and allocyclic changes.

Due to their unique geological location at the interface between land and water with high deposition rates, the evolution of a delta is often coupled with the formation of oil and coal reservoirs [110; 32; 37; 7]. The continental shelves presently supply nearly one fifth of the total oil and gas needs of the world. The first field studies investigating the geological structure of the Mississippi Delta, for example, were carried out at the beginning of the 20th century, primarily by oil companies and the US. Army Corps of Engineers [63; 64; 87; 34; 36].

The Gulf of Mexico at the mouth of the Mississippi is an important offshore oil site in the United States [110] and understanding of the mechanisms responsible for the development and distribution of deltaic deposits is therefore essential for efficient exploration [12; 37]. The exploitation of petroleum resources beneath the continental shelf began in the US in the late 1900s, first in California and later in the Gulf of Mexico. The discovery of submarine oil seeps stimulated early offshore exploitation. The first offshore well in Louisiana was drilled in 1933, while in Texas drilling started in 1936 [110]. As the seeward slope of the Gulf Coast States is extremely flat the shelf region is merely a submarine continuation of the extremely gently sloping coastal plain. Most of these early oil fields were primarily exploited from land, piers, or submersible barges, such as had been used in the coastal marshes and shallow waters. After World War II active geophysical exploration of the offshore oil fields began [110].

In recent years the study of changes in deltaic topography have come into focus. Coastal land loss due to the rising sea level combined with extreme weather events causes significant damage. Natural and man-produced subsidence may have devastating consequences for deltaic lowlands, just as damming of distributaries may lead to problems of accelerated erosion downstream. Many deltas have been turned into chemical dumps with disastrous consequences for the ecology of adjacent coastal regions. Thus, the understanding of the different aspects in the delta formation process and their interaction is not only important from an economic point of view but also for ecological preservation [70].

To get a deeper understanding of the dynamic processes involved in delta formation, laboratory experiments have been set up in recent years for quantitative measurements of sedimentation and erosion [80; 48; 8; 9; 167; 121; 123]. Delta formation experiments have been carried out in the "eXperimental EarthScape" (XES) project of the St. Anthony Falls Laboratory [85; 143; 120; 138] and in the Sheet's lab at Exxon Mobile [76]. Results have shown that the cohesion of the sediment, transported by the stream, is an essential factor in the formation of elongated deltas like the Mississippi. Due to the cohesion of the sediment, channel beds are stabilized resulting in more distinct channel patterns and less channel migration [76]. Since Devonian times, this channelization increases in complexity due to riparian vegetation which stabilizes the bed at the river banks and bars [76].

Although the techniques for topographic measurements and experimental design have advanced considerably in the last decades, computational modeling has proved to be very difficult as the systems are highly complex and large time scales have to be taken into account. Typical models combine hydrodynamics derived from the Navier-Stokes equations with an empirical erosion law based on the bottom shear stress. Sediment is transported by an advection-diffusion equation or bedload transport equations. This set of partial differential equations is then integrated using finite-element or finite-volume techniques. Examples using shallow water equations are SisBaHiA or CCHE2D [3] and extensions to three dimensions CCHE3D [4] and the Delft-3D model [57]. An overview of the most popular classical models is given in chapter 3.1.

Although classical physical models based on partial differential equations describe the details of the flow, numerical simulations of realistic river basins and catchment areas over geological periods are far beyond today's computational power. Usually these models cover small river sections over several months or years. By contrast, our knowledge of the topography and channel dynamics is traditionally derived from digital elevation measurements (DEM) and sediment records covering scales of broadly $10^0 - 10^6$ meters and $10^{-1} - 10^5$ years. Process models are required that can accommodate the changing boundary conditions governing land-surface changes and mass fluxes over these scales. Thus the challenge of geomorphological modeling is to reduce the complexity of the physical equations without modifying the characteristic behavior of the system [26; 42]. During recent years "reduced complexity models" based on the idea of cellular automata [162] have proven to be very successful in modeling the evolution of geophysical processes [26; 120; 46; 137]. Recent advances in this field have sought to overcome the limitations of classical models through the development of novel efficient cellular discretization methods and increasing reliance on high quality topographic data [26; 53]. Examples for these types of models are CEASAR [44; 45; 153] and EROS [49; 47] for river channel dynamics and alluvial sediment transport or LISFLOOD [154] for modeling flood plain dynamics; delta simulation models for river deltas have been proposed by [118, 142]. These models are based on simplified equations to obtain efficient descriptions of the landscape altering processes. But the simplifications introduce a new set of problems as the equations involved are often based on empirical descriptions instead of previously well-understood physical properties and variables. Additional complexity then emerges due to the fact that the nature of these new parameterizations may themselves be both scale and grid dependent and not easily transferable to real scales [26]. The 1994 work of Murray and Paola on braided rivers [112] can be seen as the starting point of reduced complexity modeling in fluvial geomorphology. This model simulates the generic characteristics of braided river streams in a topographically driven steady state flow system. A simple power law sedimentation-/erosion rule couples the flow with the topography change. The motivation for this type of modeling is not to simulate the deterministic evolution of a given river, but to identify the essential physics of the underlying processes. The results of these simulations then can be compared and validated with field measurements and experiments. Some prominent examples of this type of model are presented in Chap.3.2.

Based on the ideas of cellular automata, a new model is developed in this work describing the long-term behavior of the delta evolution over geological time scales. The model details are described in chapter 4. The novel aspects of the presented work lie in three fundamental points: First, the model is capable of simulating the rich dynamics related to the switching of the mouth of the river delta over periods from hundreds to thousands of years. By modifying the model parameters, the different delta types could be reproduced, according to the classification scheme of Galloway [67] (see Chap. 2.2). Second, the new reduced-complexity model is implemented to simulate the long-term development of a real delta, like the Balize Lobe of the Mississippi. Third, we rescale and interpret the parameters of the model with meaningful measured physical variables of the Mississippi, such as water and sediment fluxes, the size of the delta, observed erosion and deposition rates. This step is innovative, because reduced-complexity models are not commonly adapted to real world cases, and the physical meaning of their parameters is most often left unexplored. This step also allows us to make statements about the internal consistency of the process parameterizations (Chap.4.5).

In Chap.5 the model is extended to simulate the generation of inland deltas, using the example of the Okavango River. This type of delta does not reach the sea, but loses all its water already on the way to the ocean due to infiltration and evaporation. The Okavango model is accompanied by an experiment setup, modeling the formation of inland delta patterns on laboratory scale.

Chapter 2

River Deltas

2.1 Dynamics of River Delta Growth

River deltas form at the transition between the terrestrial and aquatic environment. Every delta has a subaqueous (delta slope) and a subaerial component (deltaic plain) even though the relative areas of these may vary considerably. The subaerial part is that portion of the delta plain above the low tide limit. In most cases the delta top and delta front deposits result in a relatively thin veneer overgrowing a much thicker layer of ancient subaqueous sedimentary sequences collectively named prodelta. The different components of the deltaic plain are shown in Fig. 2.1 and Fig. 2.2 and Fig. 2.3 shows the deposition at the river mouth.

The entire system progrades or retrogrades according to the controlling factors climate, subsidence/uplift, and sea level change. Cyclical changes in environmental conditions lead to the formation of stratigraphic layers described by the concept of sequence stratigraphy in an integrated framework. Over the past 15 years, this approach to stratigraphic analysis has been preferred by geoscientists because it ties together observations from many disciplines [28]. Sequence stratigraphy developed as an interdisciplinary method that blended both autogenic (i.e., from within the system) and allogeneic (i.e. from outside the system) processes into a unified model to explain the evolution and stratigraphic architecture of sedimentary basins [106].

The delta top unit is divided into a fluvial and a tide influenced sector. The fluvial delta starts at the seaward end of the valley and characteristically shows diverging distributary channels. Where these channels enter the sea, the bed and suspended load of fine sand and silt is eventually deposited

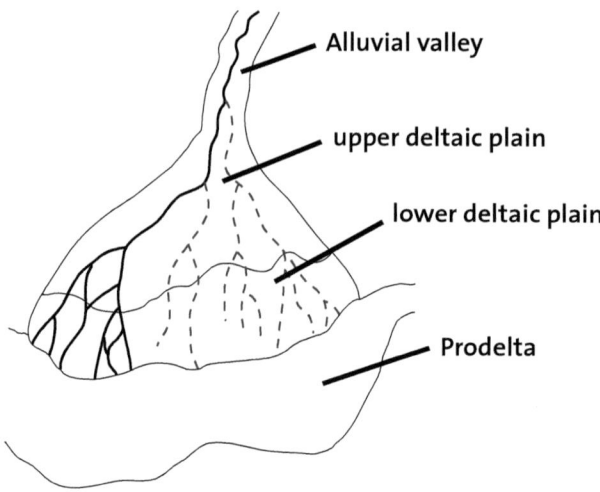

Fig. 2.1: *Components of the deltaic plain: The alluvial valley ends in the upper deltaic plain where the influence of wave and tide forces are low. The lower delta plain is that part of the delta where the riverine processes strongly interact with marine forces and wave and tidal processes are dominant. The subaqueous part of the delta is often referred to as the prodelta and forms the base of the whole delta lobe. Abandoned distributary channels are marked with dashed lines and active channels are drawn in black. The figure has been created after [32].*

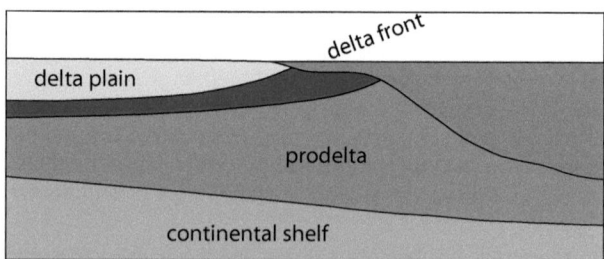

Fig. 2.2: *Cut through the delta lobe with the different components: Deltaic plain, delta front and prodelta lie above the old continental shelf.*

2.1. DYNAMICS OF RIVER DELTA GROWTH

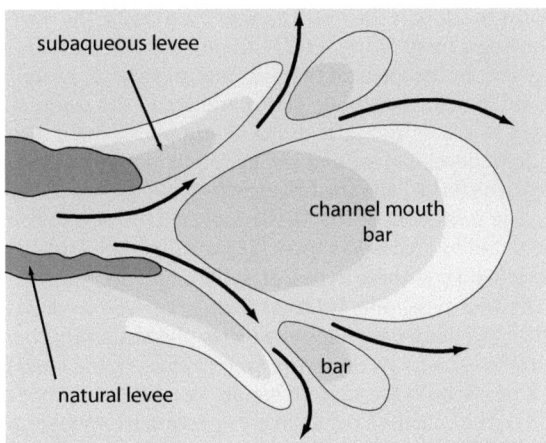

Fig. 2.3: *Schematic view of the deposits at the mouth of the delta. The main channel is confined between its subaerial levees which continue below the water level. Channel mouth bars and lateral bars form where the stream divides.*

in the form of lens-shaped sedimentary bodies forming part of the delta front. In the tide influenced sector distributary channels maintain their course developing marked levees stabilized by cohesion between finer particle deposits on the way to the mouth bar. The areas between the distributary channels are dominated by tidal channels commonly not interconnected with the fluvial system. Transport in the tidal channels is divergent and salinity is mostly normal marine depending on the rainfall upon the deltaic plain. Tide-dominated areas may deeply penetrate into the deltaic plain, especially in areas abandoned by actual flow distribution. Geometrically, the prodelta is the base segment of a half-cone. In cross-section, it is roughly triangular and hosts most of the long-term suspension of fine silt, clay and organic material. Coarse-grained intercalations imported during storms ("tempestites") and compaction induce ground failure resulting in more or less voluminous submarine landslides are quite common.

The delta plain can also be subdivided into active and abandoned zones. The active zone is that part of the delta, where the main channel splits into distributaries.

Over time, a delta can build out so far into the ocean that its gradient becomes too low to support the river's current. As a result, the river changes its course following a more efficient route to the sea. This process causes a corresponding shift in the locus of the river mouth sedimentation also known as an avulsion or delta switching. The abandoned delta eventually subsides and is eroded by waves and tides. Delta or lobe switching is a characteristic feature of delta development and can be found in many deltas such as the Danube [119], the Po [41] and the Lingayen Delta in the Northwestern Philippines [97]. The switching of the Mississippi Delta is well documented and analyzed in detail by Kolb & van Lopik [88] and Coleman [36].

In the literature [32], three types of switching mechanisms are distinguished: The first type, referred to as Switching Type I, consists of a lobe shift in which the delta progrades in a series of distributary channels. After time, the stream abandons the entire system close to the head of the delta and forms a new lobe in an adjacent region. Very often this lobe occupies an indentation in the coastline between previously existing lobes so that, over time the sediment layers overlap each other. One can find this type of delta switching in areas where the offshore slope is extremely low and the tidal and wave forces are too small for reworking the lobe [32; 163; 67]. In many cases the delta lobes merge forming major sheet-type sand banks. A Type II switching occurs when the river shifts far upstream taking a completely new course to the sea. On its way to the ocean the river stream forms a new delta lobe.

The Type III delta switching is also known as an alternate channel extension [32]. Only the dominance of sediment flux in one or more distributaries changes with time. As a result, the actual active channel will rapidly prograde seaward, while the other channel silts up. At some point, the slope of the main active channel decreases and the discharge then seeks one of the shorter distributaries. With the increased sediment flux downstream, the new channel rapidly progrades into the sea. This switching process is repeated several times forming a delta front composed of multiple beach ridges. While the lobe switching occurs over thousands of years, small sublobes are formed in the interim due to crevassing. The crevasse can be simply defined as "a break in the levee or other stream embankment" (Am. Geol. Inst. 1962). These breaks occur during river flood stages when the flow gradient advances over the natural levees. Many crevasses remain open until the low-river stage, when flow abates, and the breaks commonly seal rapidly. Other crevasses remain also open even during low-river stages resulting in rapid filling of adjacent bays. Most of the subaerial accretion in the delta can be directly attributed to crevasse deposits. During the initial stage of channel progradation land formation is not apparent. After a relatively short period of subaqueous growth, secondary channels begin to form and mouth bars aggrade to mean water level, resulting in muddy islands. Gradient advantage over the main channel is at a maximum during the early stages of growth, thus allowing extremely rapid land formation. As the original channels advance, bifurcation becomes the dominant pattern. Some distributaries remain active through much of the subdelta's life, but most are blocked with sediment at their heads after a short period of activity and are abandoned for more seaward branches. Overall deterioration of crevasse deposits begins when subdelta distributaries have lengthened considerably, thereby decreasing their gradient advantage over the major pass. The distributaries become less efficient, allowing less water to be carried by the crevasse system and causing the channel to shoal. Typically, only one or two main channels carry the majority of the flow during the final stage of the subdelta's growth cycle. These channels advance slowly seaward, while the remainder of the crevasse system subsides and deteriorates. Subsidence of the natural levees and interlevee basins by compaction of underlying unconsolidated prodeltaic clays results in rapid enlargement of ponds and lakes within the subdelta. Soon only portions of the subdelta's distributary levees that remain above water indicate the former existence of a major land feature in the shallow interdistributary embayment. Both the growth and the destruction stages of the crevasse cycle are extremely rapid when compared with major distributary progradation and deterioration in the modern delta. Scruton, Coleman and Morgan have studied subsurface characteristics of crevasse sedimentation in the modern Mississippi River Delta in detail [135; 36; 33; 110]. The concept of auto-cyclic subdeltaic sedimentation in the subsiding area has been developed from these studies. After progradation, abandonment, and subsidence of the initial subdelta lobe, a flow gradient advantage is again available in the area. Crevassing of a major distributary natural levee allows a second subdelta to prograde over previous deposits. When distributaries of the second subdelta are no longer efficient, abandonment once again allows deterioration of the deposit. Thus a layering of subdeltaic sequences is possible in an area that is subsiding rapidly. This concept is equally applicable to entire deltaic complexes.

2.2 Classification of River Deltas

The morphology and sedimentary sequences of a delta depend on climate, discharge, and sediment load as well as on the relative magnitudes of tides, waves, currents, subsidence or uplift, and sea-level change[32]. Grain size and the water depth are crucial for the shape of deltaic patterns [39; 32; 21; 117]. This complex interaction of different processes and conditions results in a large variety of surface and subsurface patterns which differ according to the local situation.

Factors controlling the morphography of deltas have been identified early by [18, 19]. Wright and Coleman [163] attempted to quantify the relations between delta morphology controlling factors, particularly wave energy fluxes. However the understanding of the full range of stratigraphic variability requires a synthesis of all sources of energy impinging on the deltaic shoreface and determination of their interactions with sediment input.

Wright and Coleman [39; 32; 163; 38] concluded that deltas result from a large variety of interacting dynamic processes (climate, hydrologic characteristics, wave energy, tidal action, etc.) sorting and dispersing sediment. By comparing sixteen and later thirty-four deltas, they found that the Mississippi Delta is dominated by the sediment supply of the river while the Senegal Delta or the São Francisco River Delta are mainly dominated by reworking through waves [82; 39; 163]. High tides and strong tidal currents are the dominant forces forming the Fly River Delta. Therefore Galloway [67] introduced a classification scheme where three main types of deltas are distinguished according to the dominant forces at work: *river-*, *wave-* and *tide*-dominated, represented by the Plaquemines-Modern Mississippi (river-dominated), the Sao Francisco (wave-dominated) and the Fly delta (tide-dominated), respectively. All marine deltas can be plotted within the triangle formed by these three examples. Each end member is characterized by a distinctive association of sedimentary facies. This simple classification scheme was later extended [21; 117; 164] to include grain size and other characteristics. The style of deltaic progradation is extremely sensitive to the variation of the relative intensities of marine and fluvial processes. As the delta progrades under the control of climate, subsidence/uplift and sea-level change the relative intensities of processes will vary and the architecture of the sedimentary units composing a delta system reflects this variation, as do the corresponding morphographic features.

2.2. CLASSIFICATION OF RIVER DELTAS

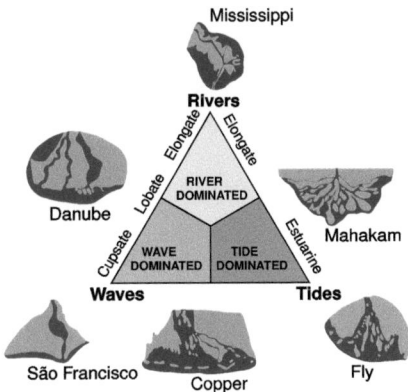

Fig. 2.4: *Classification triangle for river deltas according to Galloway [67] in river-, wave- and tide-dominated deltas. Prominent end members are the Mississippi on the river and the São Francisco Delta on the wave dominated end. A typical tide dominated end member is the Fly River Delta in Papua New Guinea.*

2.2.1 River-dominated Deltas

Fig. 2.5: *Schematic description of a highly river-dominated delta prograding through levee-constrained channels forming the typical birdfoot shape. Figure after [27]*

River-dominated deltas are indented and have more distributaries with marshes, bays, or mud flats in the interdistributary regions than other delta types. They occur when the the river and the resulting sediment transport are strong and other factors such as reworking by waves or tides are minor [32; 163]. These deltas tend to form big delta lobes into the sea with a main channel and some distributaries whose levees are exposed above sea level. When more of the flood plain between the individual distributary channels is exposed above sea level, the delta develops a lobate shape. In other cases, the river deposits its sediment more along the distributaries and protrudes out into the receiving basin forming

Fig. 2.6: *Drainage area of the Mississippi River. Figure after [32]*

finger like structures. Due to the similarity with a bird's foot, these deposits are often referred to as "bird foot deltas". The most prominent example is the modern Mississippi River Delta [32].

2.2.1.1 The Mississippi

The Mississippi River is the largest river on the North American continent, draining more than 3.2 million square kilometers, see Fig.2.6. Mean annual inflow into the delta is estimated at about $\overline{Q}$=17,000m^3/s with relatively low interannual variability [111].

The Mississippi River Delta is the modern area of land built up over the past 5,000 years by sediment deposited as the river slows down and enters the Gulf of Mexico. The deltaic process has caused the coastline of Southern Louisiana to advance gulfward from $24km$ to $80km$ currently covering a total area of about 23,000km^2. Marshes and open bays form the bulk of the delta plain. Due to the extremely low shoreline wave power (averaging only 0.03×10$^7 ergs/s$) and low tidal ranges below 1.5m, riverine processes dominate and the river mouth protrudes out into the receiving basin in the form of long finger-like configurations. The width of the main river at Tarbert Landing at the top of the delta is about 1,000m and it increases

2.2. CLASSIFICATION OF RIVER DELTAS

to about 1,400m at the Head of Passes where smaller channels spread onto the lobe and the continental shelf. At this juncture bed load composition has changed in response to the varying sediment load into finer fractions consisting predominantly of clay, silt, and fine sand. Beach ridges and other reworked marine sand bodies are relatively rare and exist only in the immediate vicinity of the abandoned river mouths. The bulk of the delta plain deposits consists of interdistributary bay deposits and crevasse or overbank splays filling brackish water depressions. The main channel is confined between two distinct levees and crevassing occurs throughout the course of the distributaries permitting tongues of sediment to extend into swamps and marshes far beyond the normal toe position of the levee. The planform of the Mississippi today is the product of repeating deltaic cycles; erosion and deposition acting together over long scales form a fractal-like coastline with numerous small islands and bays. Fisk [62] discovered that the locus of the Mississippi was shifting over time. Using radio-carbon datings he identified six major lobes.

The youngest cycle of the delta formation can be traced back to the Pleistocene, when a large amount of ocean water was sequestered in glaciers. The sea level was about 120m below the present level causing the mouth of the Mississippi to be located further out within the Gulf of Mexico. The glaciers began to melt about 10,000 years ago, and the sea level rose until approximately 5,000 years ago when it stabilized, and the formation of modern deltas began. Lake Pontchartrain was formed during the evolution of two separate delta lobes: the Cocodrie lobe expanded over the area where New Orleans presently lies, forming the lake's southern shore 4000-3800 years ago and the St. Bernard lobe completed the lake's eastern shoreline around 2,800-2,600 years ago [152]. The Lafourche delta lobe (Fig.2.7, blue) is the youngest abandoned lobe. It reached south of where Venice, LA (USA) is today. The active birdfoot or Balize delta lobe is shown in cream on Fig.2.7. A satelite image of the current Balize Lobe is shown in Fig. 2.8.

Within these major lobes Penland [126] and Roberts [129] found several smaller sub-lobes. Further investigations on the Mississippi Delta have been carried out [135; 127; 65; 35; 34; 88; 38]. Recent research has revealed that sediment load gradually has shifted from the Mississippi towards the Atchafalaya River which now carries around 86×10^6 tons of sediment per year in comparison to 150×10^6 tons of the main Mississippi stream [55].

Fig. 2.7: *Main lobes of the Mississippi. The current active Balize lobe is marked in cream; The other lobes are Sale Cypremort (>4600 B.C., orange) Cocodrie (4600-3500 B.C., violet), Teche (3500-2800 B.C., dark green), St. Bernard (2800-1000 B.C., light green), Lafourche (1000-300 B.C., light blue) and the Plaquemine delta lobe (750-300 B.C., darkblue). Figure after [32; 87]*

2.2. CLASSIFICATION OF RIVER DELTAS

Fig. 2.8: *Satellite image of the currently active lobe of the Mississippi Delta. The sediment is supplied to the sea through the elongated distributary channels of the "birdfoot" (©NASA, Earth from space collection).*

2.2.2 Wave-dominated Deltas

Many of the modern major deltas are formed in bodies of water large enough for various types of surface waves. The main role of the waves is to sort and redistribute sediment delivered by the river forming berms, beaches, barriers and spits. The geometry of many deltaic sand bodies depends not only on the magnitude and distribution of the wave forces but also on the capacity of the river to supply sediment. In rivers particle load is commonly poorly-sorted whereas wave deposits are well-sorted resulting in a higher quartz concentration. Sand bodies produced by riverine sedimentation are normally oriented at high angles to the shoreline while those produced primarily by wave action are oriented parallel to the shoreline trend. Wave-dominated shorelines are

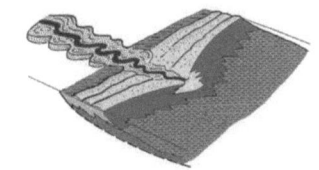

Fig. 2.9: *Schematic description of a wave-dominated delta. The sediments are redistributed along the coast and the delta forms a cuspal protrusion into the ocean. Figure after [27]*

more regular assuming the form of gentle arc-shaped protrusions. Beach ridges are common (e.g., the Nile or Niger Delta [6; 116]). Here the breaking waves cause a mixing of fresh and salt water so that the stream current immediately loses its inertia and deposits almost its entire load along the coast.

2.2.2.1 The São Francisco Delta

In Galloway's classification the São Francisco Delta is the prototype of a wave-dominated delta. The São Francisco River is the second largest river system in Brazil draining a basin of $602{,}310 km^2$ within the humid tropics. The flow is continuously high without seasonal variation providing an average discharge of 3,400 m^3/s [32]. The tropical river basin experiences extreme chemical and biological weathering. The river, therefore, transports an extremely high quantity of dilute suspended matter consisting of fine-grained sediment with an average discharge of $6 \times 10^6 t/yr$. During floods the sediment concentration within the São Francisco River equals or exceeds even that of the Mississippi. The river empties into the Atlantic Ocean and forms a triangular delta of some $735 km^2$. The bulk of the deltaic plain consists of beach ridge and dune deposits. The tidal range is less than $2.5m$ even during spring tide. Only 10% of the delta plain is subjected to tidal inundation, so the bulk of the deposits are not affected by tidal reworking. The Sao Francisco Delta displays a smooth delta shoreline with only a minor protrusion at the river mouth. The offshore slope is steep, with an average grade of 11.2%. As a result, open-ocean swells generated in the southern Atlantic are not abated and the delta shoreline is reworked by extremely high wave energy. The average wave power is $30.4 \times 10^7 ergs/sec$ compared to $0.034 \times 10^7 ergs/sec$ in the case of the Mississippi [32]. This means, that the wave energy expended in 10 hours on the São Francisco shoreline exceeds that of the Mississippi in one year. The considerable and persistent shoreline wave energy tends to produce a smooth shoreline configuration. The river mouth is highly consistent and the powerful waves attempt to seal off this particular juncture. The strong wave activity also contributes to maintaining a single distributary channel cutting across the deltaic plain. The sands consist of clean, well-sorted quartz and within the interior delta plain small swales are occasionally observed. Sandy clays within scattered organics and occasionally thin evaporite deposits are concentrated in the interior part of the deltaic plain. These deposits rarely exceed a few meters in thickness. Mouthbar deposits form the major unit of the prograding deltaic facies. These delta front deposits consist of sandy clays that accumulate

2.2. CLASSIFICATION OF RIVER DELTAS

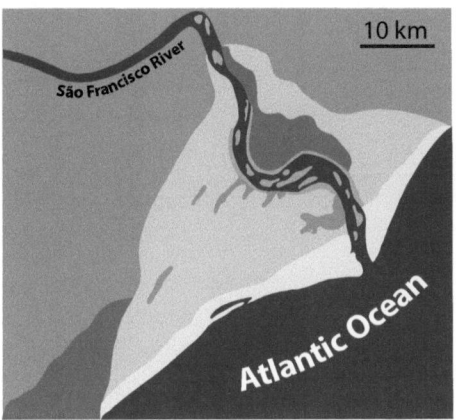

Fig. 2.10: *Map of the different facies of the São Francisco River Delta. Colors: indicate channel deposits ■, beach ridges ■, eolian dunes ■, marsh-mangroves ■, floodplain ■ hinterland ■. The figure was generated after [32]*.

beyond the immediate river mouth. Beach deposits vary from a couple of meters to $10m$ in thickness and are composed of well-sorted, clean quartz sands. The entire delta plain of the São Francisco River Delta consists of ancient shoreline deposits overlain or capped by eolian dunes. As the influence of the waves is extremely high, a complex interarchitecture such as found in the Mississippi Delta can not develop. Rather the delta top consists of monotonically accreting belts of shoreline sands, delta front sands and silts are comparatively thick, but the prodelta is underfed in suspension as the result of the strong currents carrying the fines particles away from the river mouth.

2.2.3 Tide-dominated Deltas

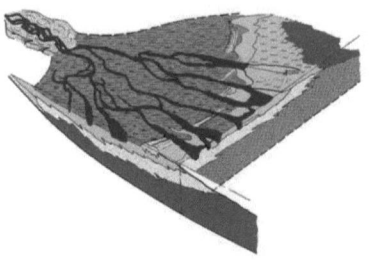

Fig. 2.11: *Schematic description of the depositional pattern of a highly tide-dominated delta. The streams divide and rejoin forming elongated sand ridges parallel to the tidal flow. Figure after [27]*

Tide-dominated deltas occur in locations of large tidal ranges or strong tidal currents. Such a delta often looks like an estuarine bay filled with many elongated islands parallel to the main tidal flow and perpendicular to the shore line. Examples are the deltas of the Ord (Australia), the Shatt-al-Arab (Iraq), the Amazon, the Ganges-Brahmaputra and the Yangtze. Turbulence produced by tidal currents inhibits vertical density stratification so that the effects of buoyancy at the river mouth are negligible. The tidal cycles cause a periodic bidirectional sediment transport leading to elongated structures parallel to the tidal flow. It is important to stress the fact that this class of delta has two types of channel networks which develop almost independently and have very few interactions: 1) riverine channels diverging towards the sea and 2) tidal channels diverging towards the land. As a result extreme contrasts in salinity can be observed between fluvial and tidal channels. Tidal currents resuspend the sediment supplied by the river reworking it into linear subaqueous sand ridges. These ridges have been described by Off [115] and Wright [165; 166]. Tidal processes, which to a large extent have strong influence on cross bedding of channel sands, cause the channels to be sand-filled in high tidal regions and lead to the accumulation of large, linear sandy tidal ridges seaward of the river's mouth. The migration of tidal channels leads to accumulation in the interdistributary regions. In the tidal system sediment accumulation works conversely to the fluvial system: fines accumulate landward and coarser grains migrate towards the sea. During the slack-water stage, high tides accumulate mud upon the flats at the upper end of tidal channels. During the falling stage and rising stages, high flow velocities are observed especially at the seaward reaches of the main channels producing long sand bars reaching beyond the geographic mouth of the river. The tidal waves also affect the channels dominated by river flow. In the main channel of the Amazon River, for example the rising tide produces a

2.2. CLASSIFICATION OF RIVER DELTAS

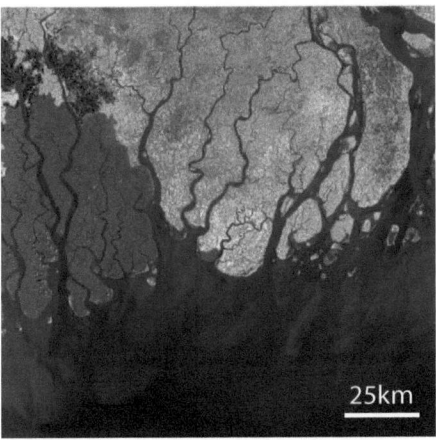

Fig. 2.12: *Landsat satellite picture of the Ganges-Brahmaputra Delta complex. The Brahmaputra Delta is one of the largest deltas in the world ans shows the typical dentrite structure of a tide-dominated delta. (Figure from USGS, "Our Earth as Art")*

standing bore wave of considerable height (max $8m$) which slowly advances landward and dissipates only at a distance of roughly $100km$. River flow is slowed down at high tide and the dammed waters greatly increase transport capacity during the falling stage favoring net export of sediment onto the continental margin. Estuarine deltas commonly possess huge prodeltas but in contrast to estuaries they maintain a certain portion of the total sediment within the delta plain. This portion is overtaken by the independent tidal channel system and redistributed towards the land.

2.2.3.1 The Fly River Delta

The model tide-dominated delta in the classification scheme of Galloway is the delta of the Fly River in Papua New Guinea. Its tributaries head in the fold and thrust belt of New Guinea. The basin receives up to 10m/year of rainfall in the uplands and still exceeds $2m/yr$ in the lowlands. About $6m$ above sea-level the Strickland joins the Fly increasing the discharge delivered to the delta by a factor of 2.5. At the mouth the total water discharge amounts to $7,000 m^3/s$. Although the drainage area is only about $76,000 km^2$, the Fly river carries a total sediment load of 85 Mt/yr [91; 92], which is greater than the combined sediment discharge of all rivers draining

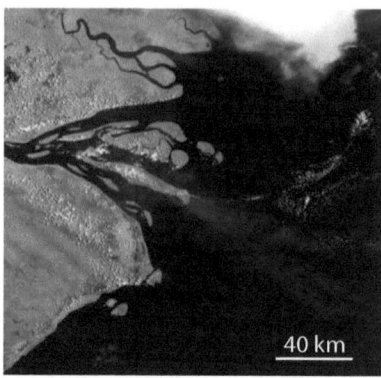

Fig. 2.13: *Satellite image of the mouth of the Fly River Delta in Papua New Guinea. Figure (©Earth Scan Lab, Louisiana State University*

the Australian continent. Mining activities in the last decade increased the discharge to $125 Mt/yr$. The Middle Fly floodplain shifts progressively from rainforest dominated to swamp, grass covered areas [51].
In the Gulf of Papua wave activities are quite low which makes the Fly an ideal example of a tide-dominated delta, whose estuary has the classic funnel-shaped geometry of a tide dominated system. Tides in the Gulf of Papua are of semidiurnal-mixed variety. Tidal currents flow predominantly across isobaths and are simplified within the Fly estuary. The tides range from $3.5m$ at the mouth to $5m$ at the apex [161]. Strong tidal currents preclude fine sediment deposition within the distributary channels. They are continuously reworking sediments in the estuary such that the turbidity of estuarine water is often greater than that of incoming river water. Sandy sediments are deposited as subtidal sand banks, capped by mud to form islands. The islands undergo rapid lateral erosion and accretion; shoreline migration rating up to $37m/yr$ has been measured from comparative aerial photographs [147]. The delta front facies is characterized by water depths of less than $13m$ and receives higher surface wave energy levels than offshore deep water environments. The water depths in the estuary are around $10m$ and the depths of the channels between the elongated estuarine islands seldom exceed $12m$ [140].

Chapter 3

Coastal and River Modeling

Deltas are obvious candidates for numerical modeling for several reasons. Direct experimentation on deltas is impossible because of their large spatial and temporal dimensions, but it is important to know how large engineering projects, such as dam constructions and waterway management will change the morphology of the associated delta. Delta models should help in long-term floodplain and coastal management especially considering the changing global climate. As deltaic deposits are also reservoirs for hydrocarbon resources, the understanding of sedimentation facies is extremely important for efficient exploitation. The sedimentary record of deltas is intrinsically complex, because deltaic systems are bypass zones, where only a portion of the sediment that travels down the river is retained. The record of the coastal zone is strongly modified by erosion and redeposition caused by storms, waves and fluvial incision. River channels and associated delta lobes tend to shift their course over time, leaving deposits in different locations.

Modeling has proven to be difficult due to the high complexity of the system. To simulate geological time scales the computational power needed is immense and classical hydrodynamical models cannot be applied. Typically these models are based on a continuum *ansatz* (e.g., shallow water equations) which describes the interaction of the physical laws for erosion, deposition and water flow [69; 157; 75; 89; 50; 17]. The resulting set of partial differential equations are then solved with boundary and initial conditions using classical finite element or finite volume schemes. Section 3.1 gives an overview of the different models and the approximations made to describe the flow. Unfortunately none of these continuum models are able to simulate realistic land-forms as the computational effort is much too high to reproduce the necessary resolution over realistic time scales. Therefore in recent years discrete models based on the idea of cellular automata have been proposed [162; 112; 113; 49; 43; 45]. These models consider water input on some nodes of the lattice and look for the steepest path in the landscape to distribute the flow. The sediment flow is defined as a nonlinear function of the water flow and the erosion and deposition are obtained by the difference between sediment inflow and outflow. This process is iterated to obtain the time evolution. In contrast to former presented models, such models can be calculated relatively quickly and several promising results have been obtained. However as they are only based on topography driven flow, a well-defined water level cannot be easily described with that *ansatz*. In Section 3.2 we will present some of the most popular Reduced Complexity Models (RCM) used in fluvial geomorphology.

3.1 Hydrodynamic Models

With the increase in computer capacity, numerical fluid dynamics calculations became applicable in hydraulic engineering. At present there are three classes of numerical models. These three classes are loosely characterized by their respective approaches to spatial discretization (finite difference, finite element, finite volume) and vertical coordinate treatment. These models solve the conservation laws derived from the Navier-Stokes equation. Sediment transport is described by an advection diffusion equation on the velocity field of the flow and empirical bedload-transport equations. Erosion and deposition are included through empirical constitutive laws based on the bottom shear stress of the flow. The equations of motion for water and sediment are obtained as follows: The water flow is mostly described by the constant density free surface shallow water equations, which are commonly used in the depth averaged approximation

$$\frac{\partial h}{\partial t} + \frac{\partial (\bar{u}h)}{\partial x} + \frac{\partial (\bar{v}h)}{\partial y} = 0 \quad (3.1)$$

$$\frac{\partial (\bar{u}h)}{\partial t} + \frac{\partial (\bar{u}^2 h)}{\partial x} + \frac{\partial (\bar{u}\bar{v}h)}{\partial y} - \frac{1}{\varrho}\left(\frac{\partial (h\bar{\tau}_{xx})}{\partial x} + \frac{\partial (h\bar{\tau}_{xy})}{\partial y} + \tau_{by}\right) + gh\frac{\partial \xi}{\partial x} = 0 \quad (3.2)$$

$$\frac{\partial (\bar{v}h)}{\partial t} + \frac{\partial (\bar{u}\bar{v}h)}{\partial x} + \frac{\partial (\bar{v}^2 h)}{\partial y} - \frac{1}{\varrho}\left(\frac{\partial (h\bar{\tau}_{yx})}{\partial x} + \frac{\partial (h\bar{\tau}_{yy})}{\partial y} + \tau_{by}\right) + gh\frac{\partial \xi}{\partial y} = 0 \quad (3.3)$$

where ξ is the elevation of the free water surface and h is the depth of the water column. The variables $\bar{u}$, $\bar{v}$, $\bar{\tau}_{i,j}$ are depth averaged quantities.

These equations are obtained from the Navier-Stokes equations, assuming that the flow depth is much smaller than the lateral extent, which means

3.1. HYDRODYNAMIC MODELS

that the vertical momentum transport can be neglected. In most cases, Boussinesq approximation and hydrostaticity are also assumed. The complete three dimensional flow field of the two dimensional models is obtained by assuming a logarithmic velocity profile and calculating only a depth-averaged solution. The first models of this type were implemented by the U.S. Army Corps of Engineers [145] discretizing the two dimensional depth-averaged free surface flow equations (Saint-Vernant or Shallow Water equation) on a fixed rectangular grid using finite difference approximation. Block structured and curvilinear gridding techniques were added later. Today these codes are included in commercial pre- and post processing softwares by Rockware [1] or Boss [2].

During the 1970's, two competing approaches to vertical discretization and coordinate treatment made their way into ocean modeling. These alternatives were based respectively on vertical discretization in immiscible layers ("layered" models) and on terrain-following vertical coordinates ("sigma" coordinate models). Both model classes use low-order finite difference schemes. One of the most popular models of this type is the Princeton Ocean Model by A. Blumberg [25]. Today, several examples of layered and sigma-coordinate models exist, including the Navy Layered Ocean Model NLOM [114], the MICOM [22; 23; 24]), and the Hallberg Isopycnic Model, HIM [72]. More recently, Ocean General Circulation Models (OGCM's) have been constructed which make use of more advanced and less traditional algorithmic approaches. Most importantly, models have been developed based upon Galerkin finite element schemes, e.g., the triangular finite element code QUODDY [93; 94] at Dartmouth University and the spectral finite element code SEOM [136] at Rutgers. These differ most fundamentally from earlier models in the numerical algorithms which are used to solve the equations of motion, and the use of unstructured horizontal grids. Recently, finite volume methods have also been adopted to achieve unstructured gridding (FVCOM) [30]. In addition, non-hydrostatic approximations have been considered in some recent models (FVCOM [30], MITgcm[96]). While most of these models only consider sediment transport as a tracer without changing the terrain surface, first attempts in coupling flow and topography change have been done by DELFT-3D, SEDSIM [98], CCHE2D [3; 56]. In CCHE2D the continuity equation for the sediment concentration is solved on top of a two dimensional shallow water model including a turbulence $k - \epsilon$ closure. Suspended sediment is advected by the water flow and bedload transport is modeled using Meyer-Peter's transport formula [105]. Transversal flow effects are accounted for by applying the formula of van Rijn [155]. Finally the elevation change is calculated using Extner's continuity equation for sediment

$$(1-p)\frac{\partial z_b}{\partial t} + \nu_b \left(\frac{\partial q_b}{\partial x} + \frac{\partial q_b}{\partial y}\right) = \nu_s(d-e) \qquad (3.4)$$

where p is the porosity of the bed material and q is the sediment flow. The parameters ν_s and ν_b describe the importance of suspended sediment transport and bed-load transport, respectively. The parameters d and e describe the erosion and deposition rates which are related to the given flow variables. The bedload sediment flux can be related to the flow variables via the following empirical equation [156]

$$q_b = \begin{cases} \alpha_b \left(\frac{u}{|u|} - \kappa \nabla z_b\right)(|u|-u_c r)_b^\beta & \text{if } |u| > u_c r \\ 0 & \text{else} \end{cases} \qquad (3.5)$$

where u is the flow velocity and u_{cr} is a critical threshold for incipient motion. The term $-\kappa \nabla z_b$ accounts for the downhill preference of the sediment on a slope.

The deposition rate can be described by

$$d = \gamma c_s h \qquad (3.6)$$

where γ is a constant, depending on the distribution of the sediment in the water column and the settling velocity of the grains. The parameter c_s is the volumetric concentration of the suspended sediment in the water column. The erosion rate is modeled by a velocity dependent law similar to the bedload equation

$$e = \begin{cases} (|u|-u_c r)_s^\beta & \text{if } |u| > u_c r \\ 0 & \text{else} \end{cases} \qquad (3.7)$$

Bank erosion in river bends is simulated by a bank erosion model including helical flow. The model is based on the shear forces, acting on the banks. Using this model Duan [56] studied the evolution of meandering rivers. Unfortunately, the computational complexity of these models is much too high to make accurate predictions on time scales where the erosion and sedimentation processes significantly change the topography of the coastline. Fully three dimensional simulations with hydrodynamic-topographic coupling have been carried out by [74] using ECOM-SED or by [57; 58] who applied the Delft-3D model to study the mechanics of river mouth bar formations.

3.2 Reduced Complexity Models

Water resource management requires integrated assessment of physical, biological and hydraulic functions of rivers, estuaries and catchments. However, time and space scales for such assessment - involving time scales of several decades and climatic change scenarios as well as whole catchment spatial scales - are completely different from the scales of process understanding and classical environmental modeling. These traditional techniques are based on physical equations and fail to incorporate the whole range of scales involved. Novel techniques have been invented to overcome these limitations. In recent years so-called reduced complexity models (RCMs) have become quite popular in geophysical modeling [26]. TOPMODEL [20], LISFLOOD [154; 13], CHILD [150] and CAESAR [44] are but a few that simulate runoff generation, floodplain inundation, landscape evolution and channel change, usually in the framework of a cellular model in which flow and sediment are routed by simple quasi-physical rules across regular gridded or triangular domains. The origins of such models, however, can be traced back to early landscape evolution models (LEMs) such as Kirkby's [86] two-dimensional and Ahnert's [5] three dimensional schemes. In these, the process representation which is driving the sediment transfer is parameterized purely through morphological expressions, where the controlling variables are local gradient and contributing area. Many RCMs also rely on quasi steady-state solutions of the flow domain for a representative discharge (or treat unsteady processes as a sequence of steady states) and therefore do not necessarily resolve behavior at the event timescale. The cellular automaton model introduced by Murray and Paola [112] is considered to be seminal in the recent history of reduced-complexity modeling in fluvial geomorphology. This model simulated the generic characteristics of braided rivers on the basis of a parsimonious topographically-driven, steady-state flow routing procedure and a simple stream power-related sediment transport law. The motivation for this type of modeling was to identify a set of rules that captured the "essential physics" underpinning the phenomenon of braiding. The aim therefore was, not to simulate the deterministic evolution of a given river reach, complicated by the sensitive dependence of the system trajectories on the initial condition, but rather to incorporate the minimum process representation required to generate the landscape-scale, generic, emergent properties of braided rivers. This is the reach-scale organization into bar and channel structures and the statistical properties of their dynamics, which implies that the relationship between the model output and the characteristics of a prototype should not be tested by expecting similar properties at all spacetime locations, but by identifying statistical similarities in ensemble macroscale properties, such as the distributions of geometries and timescales of channel bars. This introduces a challenge for the evaluation and application of reduced-complexity models in river and catchment management, where the aim will often be to simulate the precise evolution of a particular prototype, and where knowing the average condition of a river reach may be less important than knowing the local flow properties. For example, models based on the ideas of Murray and Paola have been used to conclude that the model adequately represents the channel bar association observed in real braided rivers. This, in turn, implies that the rules embodied in the model capture the essential physics.

However the model can also be validated by quantitative comparisons of model water depths and discharge with those of a prototype river. This may often be seen to display highly improbable and even unphysical lateral water surface gradients [149; 148; 54]. It is difficult to define criteria that permit a prioritization between two such distinct model tests.

While the above models are mainly related to river and floodplain dynamics, RCMs for delta formation are quite rare, and the results are often limited, suffering from non-physical results and grid effects. Several of them include initial given topographies from DEM data and cope with changing water levels and tides. Meijer [104] noted by comparing several models that the character of the statistical noise superimposed on the general grid topography influences the simulated river delta development more strongly than expected. Also digital elevation model (DEM) data input resolution affects erosion and sedimentation rates, because the constants used in the transport algorithms are not scale invariant [134]. Most models can deal with basin subsidence, which is imposed on the grid, although this has not always been tested. Another distinguishing criterion is the capability of the models to deal with variations in river discharge Q, sediment supply Q_s, or wave climate.

A three dimensional bulk sediment transport model has been developed by Ritchie et al. [128] to simulate coarse-grained delta sedimentation. The model is rule-based and uses a simple slope-and discharge dependent sediment transport, which includes lobe switching over the deltaic plain triggered by an artificial stochastic variable. It has been designed to test hypotheses about large scale sedimentation patterns. Ritchie and Gawthrop [68] modeled delta fan under the influence of varying subsidence rates and

3.2. REDUCED COMPLEXITY MODELS

eustatic change. Meijer's model [104] mimics the evolutions of a river system on exposed shelves under a fluctuating sea level regime with the aim to represent hundreds of kilometers and millions of years in scale. It uses a perfect sorting algorithm to break sediment down into coarse and fine fractions. The resulting stratigraphy of the model shows large coarse sediment bodies on the upper continental shelf, i.e. lowstand deltas, and cross-shelf incised river valleys, which are entrenched in fine sediment.

Another delta formation model was suggested by Sun et al [142] and describes the long term processes of fan formation. It does not describe channel processes but reproduces the long term behavior of the deposition on a coarse grained level. The model assumes a steady state sediment- and water flow Q_s and Q_W respectively. Continuity and momentum conservation are included in a one dimensional form along the channel and the Manning-Strickler relation is applied for the flow resistance. Water and sediment are routed due to the following rules in a $N \times N$ square lattice, with connection to the 9 next nearest neighboring cells. Diagonal cell neighbors are weighted with a factor of $\sqrt{2}$ similar to the original model of Murray and Paola. Water conservation requires that

$$\sum_j \text{sign}(i,j) Q_{wij} = 0 \qquad (3.8)$$

where Q_{wij} is the magnitude of the water discharge from cell i to cell j. The flow is distributed to the neighboring cells according to the local slope S_{ij} at a power of γ.

$$Q_{wij} = \left(\sum_{k \text{ with flow}} Q_{wij} \right) \frac{S_{ij}^\gamma}{\sum_{k \text{ with flow}} S_{ik}^\gamma} \qquad (3.9)$$

Consistent the Manning-Strickler formula γ is set constant to 0.5.

Once a channel network has been obtained, the sediment discharge Q_{sij} is obtained using semiempirical equations derived by Parker et al. [122]. The bed elevation change is described by Exner's equation which takes the discretized form

$$\Delta \eta_i = -\sum_j \frac{\text{sign}(i,j) Q_{sij}}{(1-\lambda)a^2} \Delta t \qquad (3.10)$$

where λ denotes the bed porosity and Δt the time increment; the parameter a^2 is the area of a single cell. An avulsion originating from a cell i out of neighboring cell j and into neighboring cell k is initiated if the following

inequality is satisfied:
$$\frac{(\eta_i - \beta H_{ij}) - \eta_k}{L_{ik}} \tag{3.11}$$
Here β denotes an order constant and η_i/η_k are the bed elevations in nodes i and k. The parameter L_{ik} stands for the distance to the neighbors and is $L_{ik} = a$ for the nearest neighbors and $L_{ik} = a\sqrt{2}$ for the diagonal neighbors. Once an avulsion occurs, a new channel is generated from the point of avulsion. In the simplest implementation this new path would follow the steepest descent, but the model of Sun allows a deviation from this path according to a random variable p_{ij} which is given by

$$p_{ij} = \frac{S_{ij} \exp\left[-\left(\frac{\delta\theta_{ij}}{\theta_o}\right)^2\right]}{\sum_{j'} S_{ij'} \exp\left[-\left(\frac{\delta\theta_{ij'}}{\theta_o}\right)^2\right]} \tag{3.12}$$

where the parameter θ_{ij} is the angle of the nodes downstream and θ_0 determines the standard deviation of the probability function. Lobe and channel shifting is induced by this parameter and thus it is not an internal feature of the model.

Chapter 4

A Delta Formation Model for Geological Time Scales

As discussed in the previous chapter a model is still missing, captures the evolution of deltas due to internal dynamics. For example, in many current RCM models lobe switching is imposed by an artificial switching probability and does not occur due to the instability dynamics of the channel formation and sedimentation.

Quantitative comparisons with real deltas and realistic topographies have not yet been investigated. Our understanding of small-scale physics governing delta evolution at the scale of single channels is well developed, but despite significant work on landscape-scale delta dynamics over the past few decades there has been little progress in developing simplified quantitative models. In particular, both physical experiments and numerical cellular models have been largely unable to produce the self-organized leveed channels and avulsions essential to large-scale delta dynamics.

While the physically based models presented in Section (3.1) include a good approximation of the underlying governing flow equations and can be easily rescaled to real scales, they are unable to capture realistic length and time scales due to the lack of sufficient computational power. On the other hand, reduced complexity models (RCMs) described in Section (3.2) define a set of physically motivated equations, that cannot be rigorously derived from the underlying first principle equations, but do capture the essential features of the governing processes. These models can capture the time and length scales needed for mesoscale modeling, but their rescaling behavior is often questionable.

In this chapter we introduce a new cellular automata based model which reproduces these characteristics for the first time in a simplified model. The model includes a simplified hydrodynamic routing of the flow which is completely different from the classical topography-driven routing schemes normally used in reduced complexity models. This novel approach allows us to capture several key aspects of the delta formation process that other models were not able to reproduce. These models are mainly based on Manning-Strickler like flow routing schemes which work well for channel flows, but cannot be applied in open water systems. The presented model is based on a set of simple physically motivated equations which are coupled in a nonlinear way to simulate the evolution and dynamics of the delta formation process. The model equations are completely deterministic; dynamic phenomena, like the lobe-switching stem from the nonlinear coupling of the equations and are not imposed from outside. Although the model is inspired by the geomorphological cellular automata models the basic ideas of the model equations originate from nonlinear resistor network models which have been successfully applied to a wide range of different physical problems [29; 73].

4.1 The Model

The model approximates the landscape on a rectangular grid with fixed spacing. A surface elevation H_i and a water level V_i are assigned to the nodes. Both H_i and V_i are measured from a common base point, which is defined by the sea level. On the bonds between two neighboring nodes i and j, the hydraulic conductivity for the water flow from node i to node j is defined as

$$\sigma_{ij} = c_\sigma \begin{cases} \dfrac{V_i + V_j}{2} - \dfrac{H_i + H_j}{2} & \text{if} > 0 \\ 0 & \text{else.} \end{cases} \tag{4.1}$$

The variable σ_{ij} can be interpreted as the average water depth on a bond between node i and node j, thus deep channels can carry more flow than shallow ones. A sketch of a distributary channel segment is shown in Fig. 4.1. The model considers only surface water flow and therefore the hydraulic conductivity is set to zero when σ_{ij} is negative. Ground water flow into an open channel is not desired in the model either, thus the conductivity is also set to zero if the water level in the source node in the flow direction is below the ground. This is applied even if σ_{ij} is larger than zero. Water is routed downhill due to a hydrostatic pressure gradient arising from the difference in the water levels in node i and node j. This pressure drop induces a flow

4.1. THE MODEL

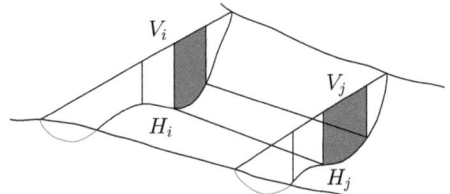

Fig. 4.1: *Cross section of a distributary channel. The height of the water level is V_i and the elevation of the channel bed H_i.*

I_{ij} between nodes i and j

$$I_{ij} = \sigma_{ij}(V_i - V_j). \qquad (4.2)$$

where σ_{ij} is the hydraulic conductivity defined in Eq.4.1. Note that this equation implies a nonlinear relation between the current I_{ij} and the waterlevel V_i as $\sigma_{ij} = \sigma_{ij}(V_i, V_j)$ is itself a function of V_i. Equations 4.1 and 4.2 resemble the Hagen-Poiseuille equation for laminar flows in pipes, where the flow is described as a product of the pressure drop and the cross-sectional area of the pipe. The flow routing due to the pressure drop is much more physical than topography driven routing schemes like Manning-Strickler, classically used in environmental engineering. Furthermore, the calculation of the square root of the slope is computationally much less efficient than the scheme described by Eqns.(4.1)-Eq.(4.2).

As the change of topography takes place on a much longer time scale than water flow, the water flow can always be assumed to be in a quasi steady-state regime. The conservation of mass of water is given by the continuity equation for the flows entering and leaving node i

$$\sum_{N.N.} I_{ij} = 0. \qquad (4.3)$$

where the sum runs over the four nearest neighbors of an inner node. The resulting system of equations then is solved using a relaxation method

$$V_i' = V_i + \delta_R \sum_{N.N.} I_{ij} \qquad (4.4)$$

where δ_R is the relaxation parameter tat has been fixed at 0.01. The relaxation technique also allows a physical transition from one steady state solution to another, which would not be possible by a direct solution of Eq.(4.3) due to the discontinuities in σ_{ij}. Boundary conditions are needed to close the system of equations. At the sea boundary the water level is kept fixed (Dirichlet boundary conditions), while inflow boundaries and noflow lateral boundaries are applied on the land. Water and sediment are injected into the domain by defining inputs I_0 of water and sediment s_0 at an entrance node (see Fig.4.2). The landscape is initialized by a given ground water table, and runoff is produced when the water level exceeds the surface of the water table.

Assume that sediment is advected instantaneously by the water flow, thus all sediment entering a node has to be distributed to the outflows according to the corresponding water outflow. For the sediment routing we therefore obtain

$$J_{ij}^{out} = \frac{\sum_k J_{ik}^{in}}{\sum_k |I_{ik}^{out}|} I_{ij}^{out}, \qquad (4.5)$$

where the upper sum runs over all inflowing sediment and the lower one over the water outflows. A sediment inflow s_0 is defined in the initial bond. The sedimentation and erosion process is modeled by a phenomenological relation, based on the flow strength I_{ij} and the local pressure gradient imposed by the difference between the water levels in the two nodes, V_i and V_j. The sedimentation/erosion rate, dS_{ij}, is defined through

$$dS_{ij} = c_1(I^\star - |I_{ij}|) + c_2(V^\star - |V_i - V_j|), \qquad (4.6)$$

where the parameters $I^\star$ and $V^\star$ are erosion thresholds, and the coefficients c_1 and c_2 determine the strength of the corresponding process. The first term $c_1(I^\star - |I_{ij}|)$ describes the dependence on the flow strength I_{ij} [160] and is widely used in geomorphology, while the second term $c_2(V^\star - |V_i - V_j|)$ relates sedimentation and erosion to the flow velocity, which in the model can be described by $I_{ij}/\sigma_{ij} \sim |V_i - V_j|$ where $|V_i - V_j|$ represents the hydrostatic pressure drop. Considering that the relation between the water level V_i and the current I_{ij} is not linear, the two terms in Eq.(4.6) are not linearly dependent.

The sedimentation rate dS_{ij} is limited by the sediment supply through J_{ij}, thus in the case that dS_{ij} exceeds J_{ij} the all of the sediment is deposited on the ground and the sediment flow J_{ij} set to zero. In the other cases J_{ij} is reduced by the sedimentation rate or increased if erosion occurs. The erosion process is also supply-limited which means that the erosion rate is not allowed to exceed a certain threshold θ; so, if $dS_{ij} < \theta$, then $dS'_{ij} = 0$. Note that in the case of erosion, dS_{ij} is negative. The threshold θ is also introduced

4.1. THE MODEL

for numerical stability reasons, so if the potential rate dS_{ij} is smaller than a given value θ, we do not allow erosion locally in that time step. This condition occurs very rarely in the simulation. For the deposition case, $(dS_{ij} > 0)$ Eqn.(4.6) gives a deposition capacity

$$dS^{(1)}_{\max} = c_1 I^* \quad \text{resp.} \quad dS^{(2)}_{\max} = c_2 V^*. \tag{4.7}$$

In summary we have the following situation:

$$dS'_{ij} = \begin{cases} dS_{ij}, & J'_{ij} = J_{ij} - dS_{ij} & \text{if} \quad \theta < dS_{ij} < J_{ij} \\ J_{ij}, & J'_{ij} = 0 & \text{if} \quad dS_{ij} > J_{ij} \\ 0, & J'_{ij} = J_{ij} & \text{if} \quad dS_{ij} < \theta \end{cases} \tag{4.8}$$

The parameter c_k, $k = 1, 2$ modulates how much sediment can be eroded or deposited. The fact that the parameter c_k is identical for both erosion and deposition is a limitation of the model because it has to capture the effect of both processes. In the erosion mode, c_k represents the erodibility of the surface, sediment size etc. while in the deposition mode, c_k represents primarily the settling velocity of the suspended particles and the trapping efficiency in a bond in relation to the flow.

The ratio between suspended load and bedload can vary immensely from river to river. According to the existing literature, the clastic load of the rivers of the world is about four times larger than the dissolved load. Similarly, the suspended load is several times greater than the bed load transport. Values commonly fall within the bounds 85-99% suspended load to 1-15% bed load. The precise proportion depends on the transport power of the river, the hydrological regime, the geological structure of the catchment and human activity[11]. For simplicity the model does not distinguish between bed and suspended load. Sediment grain size effects also are not considered. Due to erosion or deposition, the landscape is modified according to

$$H'_i = H_i + \frac{\Delta t}{2} dS'_{ij} \tag{4.9}$$

$$H'_j = H_j + \frac{\Delta t}{2} dS'_{ij}, \tag{4.10}$$

where the sediment deposits equally on both ends of the bond. The new topography is marked with H'_i. The Eqns.(4.9), (4.10) hold for both erosion and deposition and ensure sediment continuity (Exner's Equation), because the change of the landscape in Δt is given by $(H'_i - H_i)/\Delta t = 1/2 dS'_{ij}$, where dS'_{ij} is exactly the change of the sediment flux through node i. The time step

Δt refers to the erosion- and sedimentation process. Within this time, the water flow is assumed in a steady state. Iterating Eqns. (4.1-4.10) determines the time evolution of the system. Finally in a real system, water currents in the deep sea lead to a smoothing of the bottom which is modeled by the following expression

$$H'_i = (1-\epsilon)H_i + \frac{\epsilon}{4}\sum_{N.N.} H_j, \qquad (4.11)$$

where ϵ is a smoothing constant determining the strength of the smoothing process. The sum runs over all nearest neighbors of node i.

4.2 Simulation

The simulation is initialized with a valley on a rectangular $N \times N$ lattice with equally spaced gridlines. The valley runs downhill with slope S along the diagonal of the lattice and the hillside slopes of the valley increase from the bottom of the valley laterally according to a power law with exponent α. In the simulations the value of α was chosen to be 2.0. Under the sea the landscape is flat with a constant downhill slope. Furthermore, we assume the initial landscape to have a disorderly topography by assigning uniformly distributed random numbers to H_i. This variable is then smoothed out according to Eq. (4.11). The water level V_i of the system is initialized with a given ground water table. In reality, the distance from the ground water to the surface is minimal on the bottom of the valley and increases uphill. This is obtained in the simulation by choosing the water level V_i on an inclined plane δ below the bottom of the valley. The initial water table at the bottom of the valley was set to $\delta = 0.0025$ below the surface. The slope of the plane is the same as the slope of the valley S,
which also keeps the river close to the bottom of the valley. Since we are interested only in studying the pattern formation at the mouth of the river, the braiding conditions of the upper river only determine the width of the delta front. On the sea boundary of the domain the water level $H_i \leq 0$ is a constant and set to zero. This Dirichlet boundary condition allows the water to flow out of the system into the sea. On the land side we apply no flow boundary conditions except at the location where an input water flow is injected into the first node of the lattice (Neumann boundary conditions). A sketch of the initial landscape is shown in Fig. 4.2.

4.2. SIMULATION

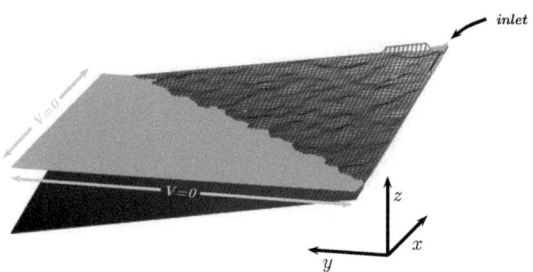

Fig. 4.2: *Sketch of the initial and boundary conditions for the simulation with the river delta model. Water and sediment are injected at the upper node (inlet) and the water levels on the sea boundaries are kept constant ($V_0 = 0$). The landscape is initialized as an inclined plane with a disordered topography on the top. The water surface (light-gray) is parallel to the horizontal xy-plane.*

An initial channel network is created by running the algorithm without sedimentation and erosion until the water flow reaches a steady state. The sedimentation and erosion procedure is then switched on and the pattern formation at the mouth of the river is studied. Simulations have been run using several different parameter configurations.

The presented model is able to reproduce the two main delta types of the Galloway classification[67], namely river- and wave-dominated deltas. Tides are not considered in the model, so that tide-dominated deltas cannot be reproduced. According to the dominance of the different processes, entirely different coastline shapes can be observed. The smoothing procedure Eq. (4.11) favors the formation of an estuary by reworking the coastline at the river mouth, smoothing out small channels and bars.

The stream-dominant erosion term $c_1(I^\star - |I_{ij}|)$ in Eq. (4.6) favors the formation of river-dominated birdfoot shaped deltas. She second term $c_2(V^\star - |V_i - V_j|)$ in Eq. (4.6), which depends on the pressure gradient represented by the height-difference of the water levels in the nodes i and j, distributes the sediments more equally when entering the sea, leading to lobate shaped delta patterns. These patterns are similar to the distributary structure of the Lena, the Selenga and the Mahakam river deltas.

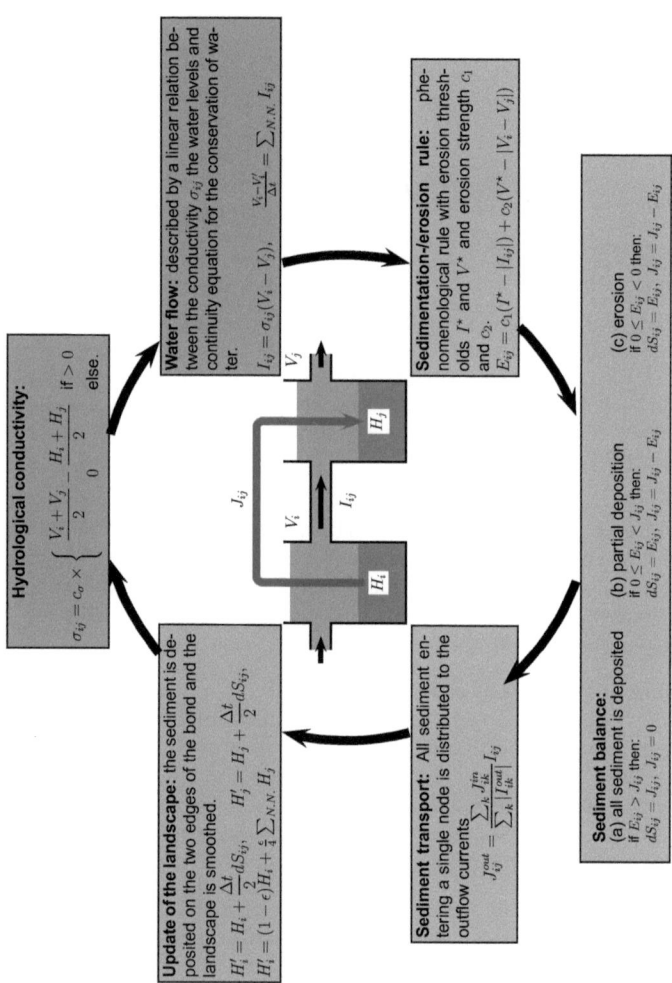

Fig. 4.3: *Schematic description of the model cycle*

4.3 Modeling different delta types

In Fig.4.4(a), a snapshot of a simulation of a birdfoot delta ($c_2 = 0$) is shown. A map of the Mississippi River is presented in Fig. 4.4(b) for comparison. In both cases the main channel prograds into the ocean, depositing sediment mainly on its levee sides forming the typical birdfoot shape. When the strength of the main channel decreases, side channels start to appear, breaking through the sidebars as can be seen in the sequence of Figs. 4.23(b) and 4.23(c). At the beginning of the delta formation process, sediment transport is equally distributed among the different channels and leads to a broader growth of the delta front along the coast. With time, the side channels are gradually abandoned and the sediment is primarily routed through the main channel, thus thus causing the main channel to grow faster than the other channels and forming the typical birdfoot shaped deposits.

Figure 4.5(a) shows another type of delta, where the smoothing action of the waves reworks the deposits at the river mouth and redistributes them along the coast. Here the river could build up only a slight protrusion in the immediate vicinity of the river mouth. Due to the redistribution of sediment, small distributaries are reworked, leaving only the main channel. The same happens in wave-dominated deltas like the São Francisco River in Brazil or the Nile River in Egypt. For comparison, a map of the São Francisco River Delta is shown in Fig. 4.5(b). Here, the coastline has been straightened by wave activity and consists almost entirely of beach ridges, that have the typical triangular shape. This flattened deposit can also be found in our simulation results. As evaporation is not considered in the simulation, small ponds and abandoned channels remain in the sedimented zone instead of disappearing with time.

Finally, if the term $c_2(V^* - |V_i - V_j|)$ dominates the sedimentation/erosion process, a half moon shaped delta with many small islands, lakes, and channels appears. This delta type shows more activity in the channel network than the others, forming a fractal like coastline. The channels split and fan out delivering sediment to a large area. When a channel is blocked due to sedimentation, the whole delta lobe switches to another location. During the simulation, the delta switching occurred several times. The switching will be described in detail in Section 4.5.

At this point, we compare the river delta patterns generated by our simulations with real structures. As a basis we analyze the self-similarity behavior between the real and simulated river deltas using the box counting algorithm [61; 151]. The box counting dimension is a quite common measure in

40 CHAPTER 4. A DELTA FORMATION MODEL

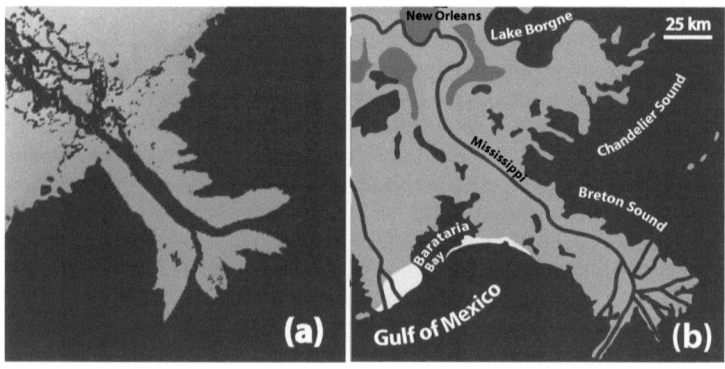

Fig. 4.4: *Comparison of a simulation of a river dominated delta (a) with the current active lobe of the Mississippi (b). The birdfoot shape of the delta can be identified clearly. Colors: channel deposits , sand ridges , swamps and marshes . Figure (b) was generated after [32].*

4.3. MODELING DIFFERENT DELTA TYPES

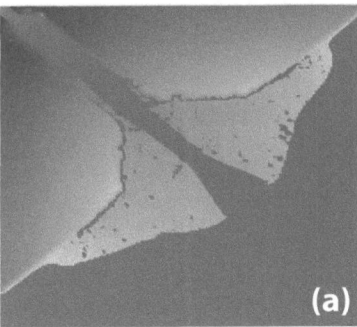

Fig. 4.5: *Simulation of a wave-dominated delta (a). While waves are shaping the coast, the river deposits sediment and supplies sand to the beaches. As the simulation does not include evaporation, ponds and inactive channels in the deposition zone do not disappear as in the map of the real river shown in (b). The parameters in the simulation were $N = 179$, $I_0 = 1.7 \times 10^{-4}$, $s_0 = 0.0015$, $c_\sigma = 8.5$, $c_1 = 0$, $c_2 = 0.1$ and $I^* = 1.3 \times 10^{-4}$. Smoothing was applied every 200 time steps with a smoothing constant $\epsilon = 0.01$. For comparison, a map of the São Francisco river delta in southern Brazil is shown (b). Colors indicate: channel deposits* ▨, *beach ridges* ▨, *eolian dunes* ▨, *marsh-mangroves* ▨, *the floodplain* ▨ *and the hinterland* ▨. *The figure was generated after [32].*

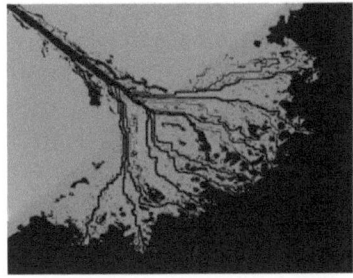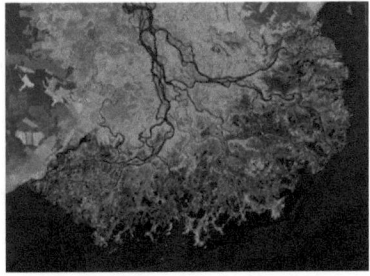

Fig. 4.6: *Comparison of a simulation of a river-dominated, lobate-shaped delta (left) with a satellite image of the Selenga Delta entering Lake Baikal in Siberia (right)*

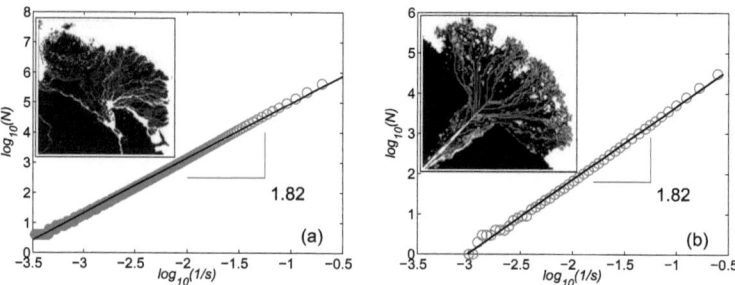

Fig. 4.7: *Figure (a) shows the scaling behavior of the Lena River Delta. On the y-axis we show the logarithm of the number of boxes $N(s)$ of size s which are necessary to cover the subaerial surface plotted versus the logarithm of the inverse box size. The straight line is a power law fit $N \sim s^{-D}$ with exponent $D = 1.82$. In the inset a satellite picture of the Lena delta is shown. (b) Scaling behavior of a lobate shaped delta simulation (c.f. Fig. 4.6) where the slope was calculated to be 1.82.*

geomorphological pattern analysis and has been used by many authors to characterize river basin patterns and coastlines [95; 130; 131]. It gives an approximation of the fractal dimension of an object in a certain scaling range. The box counting fractal dimension of a self similar object is defined by the limit

$$D = \lim_{s \to 0} \frac{\log N(s)}{\log 1/s} \quad (4.12)$$

where $N(s)$ is the minimum number of open balls of radius r which are necessary to cover the object. For the real satellite image as well as for the simulated river delta, Figs. 4.7, 4.10, and 4.9 show that the variation with the cell size s of the number of cells N covering the land follows typical power laws over more than 3 decades

$$N \sim s^{-D}, \quad (4.13)$$

where the exponent D is the fractal dimension. Moreover, the least square fit of this scaling function to the data gives exponents which are strikingly close to each other, namely $D = 1.82 \pm 0.1$ for the real Lena river delta and $D = 1.82 \pm 0.1$ for the simulation of a lobate river dominated delta. Figure 4.7 shows the scaling behavior of the simulated delta pattern in comparison with the fractal dimension of the Lena Delta. The change of the fractal dimension during the evolution of the delta simulation is shown in Fig. 4.8. After a gradual growth the delta occupies a constant space, switching the

4.3. MODELING DIFFERENT DELTA TYPES

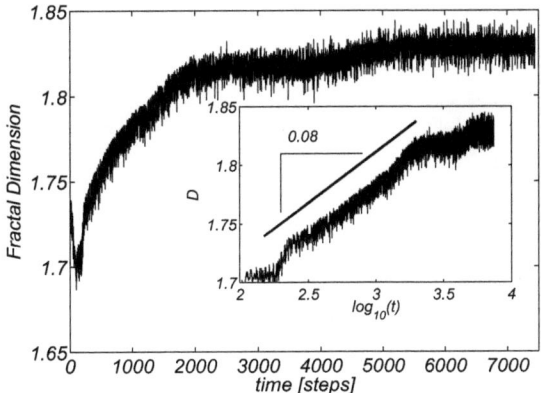

Fig. 4.8: *Change of the fractal dimension during the growth of a lobate-shaped delta (c.f. Fig. 4.6). At the beginning the delta grows exponentially with a characteristic time of 80 timesteps and then becomes saturated at a fractal dimension of 1.82. The inset shows a semilog plot of the first 4 million steps.*

delta lobes and forming new deposition sheets on top of the older lobes. For the wave-dominated delta types we obtain comparable results, namely 1.88 for the São Francisco Delta and 1.90 for the simulation of an equivalent wave dominated delta, Fig.4.9. In the case of the birdfoot of the Mississippi River we obtained a fractal dimension of 1.81±0.1 and the corresponding simulation yielded $D = 1.79$ (Fig.4.10). In contrast to lobate shaped deltas, the growth of the birdfoot is more linear, with events of larger delta growth interrupting the gradual change. Another important characteristic of tree like network structures is the branching index. The Horton-Strahler branching index was first developed in hydrology by Robert E. Horton (1945) and Arthur Newell Strahler (1952, 1957) to describe the stream size in a river network based on a hierarchy of tributaries and its branching complexity. Formally the Horton-Strahler number of a branch in a directed tree structure is defined in bottom-up order, as follows:

- The successing nodes of a node i are challed child nodes.
- If the node is a leaf (has no children), its Strahler number is one.
- If the node has one child with Strahler number i, and all other children have Strahler numbers less than i, then the Strahler number of

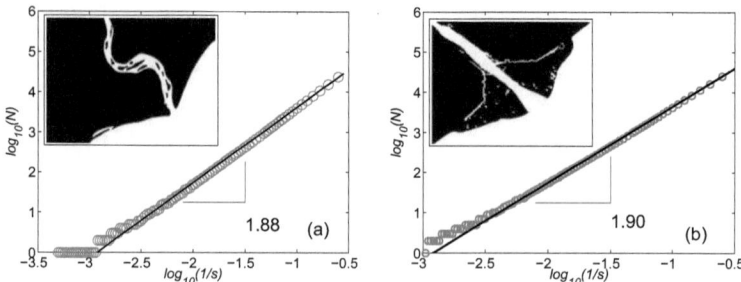

Fig. 4.9: *Comparison of the fractal dimension of a wave-dominated delta simulation. (a) Fractal scaling behavior of the São Francisco Delta with exponent of $D = 1.88$ and the fractal scaling behavior of the simulated delta where an exponent of $D = 1.90$ has been obtained (a).*

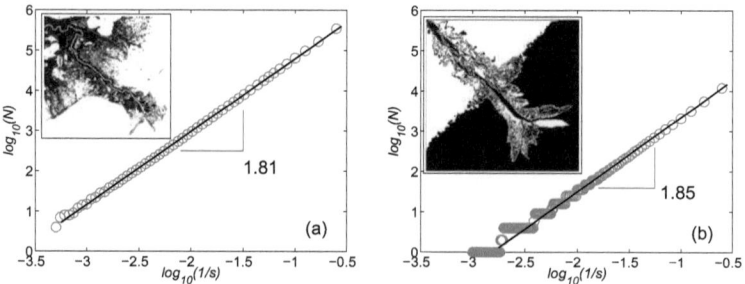

Fig. 4.10: *Comparison of the fractal dimension of the Mississippi Delta with exponent $D = 1.81$ (a) with a simulation of a birdfoot delta and exponent $D = 1.79$ (b).*

4.3. MODELING DIFFERENT DELTA TYPES

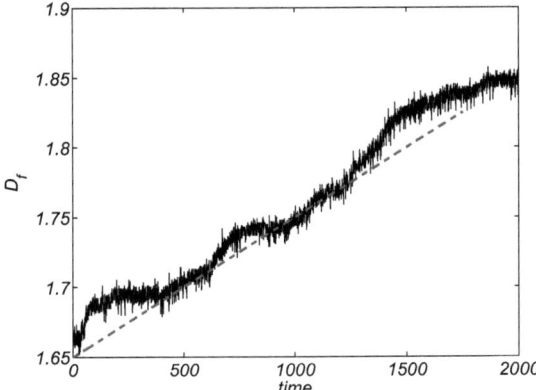

Fig. 4.11: *Growth of the fractal dimension of a birdfoot delta simulation. The growth is almost linear (dashed line) with strong events interrupting the gradual growth.*

the node is i again.

- If the node has two or more children with Strahler number i, and no children with greater numbers, then the Strahler number of the node is i + 1.

Wave-dominated deltas usually do not show a strong branching structure due to the fact that wave forces merge small distributary channels such that only the main stream remains. Thus, the branching index B of wave dominated deltas is usually 1. This is also observed in the simulation (c.f. Fig.4.5). The birdfoot delta shows a branching index of $B = 2$ where the stream divides at the end into two distributary channels, where one of these streams becomes dominant after a certain time and a new channel emerges due to a break in the levee. It is common to express the results of the Horton Strahler ordering in terms of the ratio

$$R_b = \frac{\langle N_k \rangle}{\langle N_{k-1} \rangle}. \tag{4.14}$$

Here N_k is the number of streams of order k. For the Mississippi like birdfoot delta simulations, we obtain a branching ratio of $R_b = 2$. For wave-dominated deltas with only one main stream the branching ratio is not well

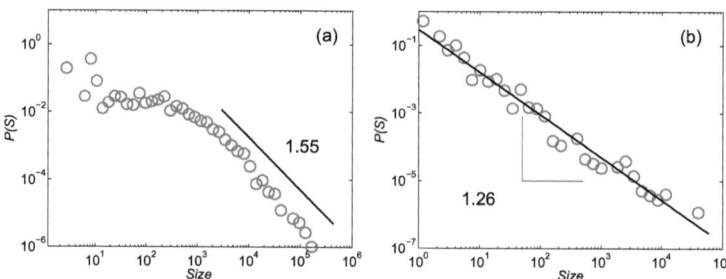

Fig. 4.12: *Size Distribution of the islands in the Lena Delta. The distribution shows a power law behavior with exponent 1.5 over almost three decades. For the simulation presented in Fig.4.6 we obtain an exponent of 1.26 ± 0.1 (b).*

defined. The Mississippi Delta has a branching ratio of $R_b = 4-5$ depending on which outlets are still considered as distributary channels. Note, that for the calculation of the branching ratio the ordering of the channels starts from the sea. In the case of lobate shaped deltas like the Lena or the Selenga Delta the classical Horton-Strahler ordering for river networks cannot be directly applied, because the channels in the delta region divide and rejoin and therefore a unique flow path is not defined. Instead the number of branches can be estimated by counting the islands and assigning one branching to each island [59]. Using a cluster algorithm, we determined the number and size of the islands in the simulation (c.f.Fig.4.6) and for the Lena Delta as well as for the Birdfoot of the Mississippi. The island size distribution for the Lena Delta is presented in Fig.4.12(a) and shows a power law behavior over two to three decades with exponent 1.5±0.1. In the case of the Mississippi Delta we obtain a value of 1.52 ± 0.1 (Fig. 4.13, which is close to the exponent obtained for the Lena Delta, while the simulations show a much smaller exponent of 1.26±0.1 (Fig. 4.12(b)). This could be due to grid resolution effects, as the simulation is performed on a much smaller grid than the DEM discretization.

For the Lena Delta and the corresponding simulation we also measured the eccentricity and orientation of the single islands. The islands are normally elongated in the flow direction and their mean eccentricity has been determined to be around 0.88. The histogram of the eccentricity is shown in Fig.4.14(a). As islands in streams are normally oriented along the main flow direction we analyzed the orientation of the islands for the Lena and the simulated delta. In the case of the Lena Delta we confirmed that there

4.4. INTERPRETATION OF THE MODEL PARAMETERS

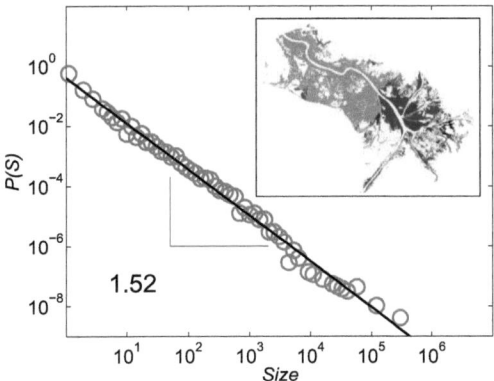

Fig. 4.13: *The distribution of the size of the islands in the Mississippi Delta shows a power law behavior over more than four decades. The exponent was determined to be 1.52 ± 0.1.*

was no preferred flow direction due to tides or coastal currents, which would lead to a non-uniform distribution of the island orientation over the delta lobe. The histogram of the orientation of the islands is plotted in the inset of Fig.4.14(a). The results of the simulation are summarized in Fig.4.14(b) where we obtained a mean eccentricity of 0.89. The orientation distribution of the islands seems also to be uniform over the delta lobe, but lacks better statistics. Also the influence of the grid is visible for the values $\varphi = 0°$ and $\varphi = 90°$.

4.4 Interpretation of the model parameters

To compare the simulation results with an existing birdfoot delta we have to reinterpret the model and given dimensions to the grid, and to the parameters by comparing the model variables with the observed delta size and water- sediment fluxes. We also need to verify whether the simple erosion-sedimentation rate in the model Eq.(4.6) provides a meaningful process description. These are necessary tests to judge whether the model is internally consistent and can provide physically correct behavior. To check whether the RCM model produces meaningful results we compared some easily accessible data, like sediment and water fluxes and the surface pattern of a compa-

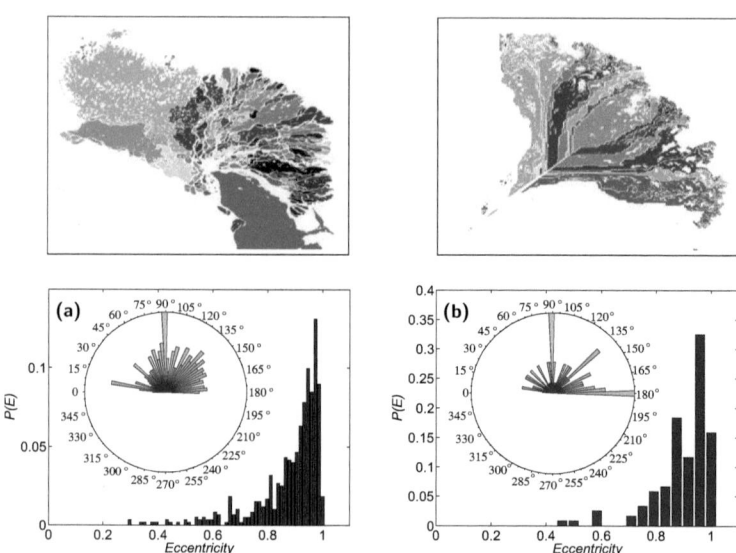

Fig. 4.14: *Figure of the different islands in the Lena Delta (a) and a corresponding simulation (b). The lower plots show the histograms of the eccentricity of the islands in the Lena Delta (a) and in the simulation (b). For the Lena Delta the mean eccentricity was found to be 0.88 and for the simulated delta 0.89. The insets show the distribution of the orientation of the islands with respect to the horizontal axis. For both, simulation and Lena Delta the distribution is almost uniform over the delta lobe, which means that there is no preferred direction of the flow.*

4.4. INTERPRETATION OF THE MODEL PARAMETERS

rable delta with the simulation, and then rescale the simulation variables. The rescaled model allows us to compare other quantities from the simulation with the real case, e.g. the volume of the deposited sediment or the growth dynamics of the delta. These predictions are essential to the rescaling proces, because they strongly depend on the internal dynamics of the RCM model and not only on the input conditions. A similar surface pattern at a given timestep, and correctly scaled inflow/outflow conditions for the sediment, do not necessarily mean that the subaqueous deposits are comparable because sediment can enter and leave the domain, and are redistributed by the erosion- sedimentation rule. In this sense, the comparison of the lobe deposits gives us the ability to judge whether the internal nonlinear dynamic is consistent with the processes in a real delta. We chose to perform the comparison on the Mississippi Delta because it is the largest river-dominated delta for which we have good records of bathymetry, water and sediment discharge, and change over time [83; 111; 40]. The proceedure entailed first collecting observations of water and sediment fluxes and estimate the volume of the Balize (birdfoot) Lobe of the Mississippi Delta and second we rescaling the horizontal and vertical resolution of the modeling grid, the time step and fluxes to match the observed data. The form and parameters of the erosion- sedimentation law in the model we interpreted to show that the model equations were internally consistent.

4.4.1 Water-sediment fluxes and lobe volume

The Mississippi River is the largest river on the North American continent, draining more than 3.2 million km^2. Mean annual water inflow into the delta in the last few centuries is estimated at about $\overline{Q} = 17000\ m^3/s$, with rather low interannual variability [111]. However, recent human activity has caused greater variability in the sediment load delivered to the head of the delta at Tarbert Landing (91.61W, 31.00N). Soil conservation, sediment trapping in reservoirs and levee construction since the 1950s have led to a reduction in the suspended sediment load from about 260 to 150 Mt/yr [83; 40; 144]. Human influence on the geomorphological processes is an uncertainty for the modeling because detailed data for the pre-settlement era is not available. Nevertheless the influence of dam and levee construction on the Mississippi in the last 50 years could have affected the last 5-10% of the total formation time of the Balize lobe. Therefore it can be assumed that human influence is not the dominant factor in the geomorphological processes in the Mississippi Delta. Since we are studying the development of the delta over time scales prior to human influence, we took the average annual sediment load into the delta, $\overline{Q_s} = 214 Mt/yr$. The width of the main river at Tarbert Landing at the top of the delta is about 1000 m and increases to about 1400 m at the Head of Passes where smaller channels spread onto the lobe and the continental shelf. There, the bed sediment composition has changed in response to the changing of the sediment load towards finer fractions. We took the data from earlier surveys which gave the median diameter $d_{50} = 0.17\, mm$ with 64% fine sand and 35% clay [83] and applied a corresponding dry bulk density $\rho = 1.4\, t/m^3$, accounting for consolidation of the sediment as it was deposited.

This study is concerned with simulating the formation of the most recent Mississippi Delta lobe, the Balize Lobe. From geological records, it is known that the Balize Lobe formed during the last 800-1000 years [55; 133; 129]. We estimated the lobe volume with two independent methods from observations of sediment inflow and directly from bathymetric data.

The first method assumes that the incoming sediment flux is constant over the entire period $\tau = 800\text{-}1000\, yr$ and equal to the long-term mean $\overline{Q_s} = 214 Mt/yr$. When we applied this flux and assume deposition with bulk density $\rho = 1.4\, t/m^3$ we got a total volume of 12.2-15.3 $\times 10^{10}\, m^3$. The study of [40] on the western flank of the Lobe has shown that approximately 40% of the delivered sediment is advected to distal regions of the shelf by current-driven resuspension and mass movements. Considering an outflow of 40% of the sediment for the Lobe area as a whole, we got an estimate of the total lobe volume

$$V_Q = 7.3 - 9.2 \times 10^{10} m^3 \qquad (4.15)$$

The second method uses the USGS Coastal Relief DEM [53] with a resolution of 3 arcsecs (cell size about 83-87 m) to determine the Lobe volume. First the original coastline was reconstructed from geological records using interpolation techniques. The continental shelf at the coast of southern Louisiana reaches about 50-60 km into the ocean, where the depth is not more than 20-50 m. It then drops immediately down to a depth of several hundred to more than one thousand meters. The bathymetry of the coast-shelf transition before the formation of the Balize Lobe was obtained by using a special interpolation technique explained as follows: Let $H(\mathbf{x})$ be the elevation of a point $\mathbf{x}$ in the Mississippi Region at present and $\tilde{H}(\mathbf{x})$ be the landscape 1000 years before today. Outside of the deposition zone we assume that the topography did not change,

$$\tilde{H}(\mathbf{x}) = H(\mathbf{x}) \qquad \forall \mathbf{x} \notin \mathcal{D}, \qquad (4.16)$$

4.4. INTERPRETATION OF THE MODEL PARAMETERS

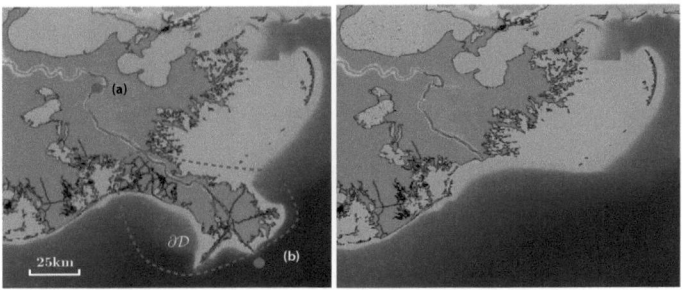

Fig. 4.15: *The topography of the coast of southern Louisiana before (right) after (right) cutting the Balize lobe. The actual course of the coastline is marked with a black line and the boundary ∂D of the erased area is marked with a red dashed line. Within the erased region the elevation of the topography was calculated by solving Laplace's equation using Dirichlet boundary conditions with the values of the present elevation on ∂D. The location of the two points which have been used for calculating the average slope are marked with red where (a) has an elevation of about +5m and (b) around -60m. Head of Passes is marked with a red cross, Tarbert Landing is outside the map.*

where $\mathcal{D}$ denotes the deposition area of the Balize Lobe (see Fig.4.15). The boundary of the domain $\mathcal{D}$ is fixed by the course of the ancient coastline based on [110; 132] and the abyss on the sea side. Values inside the domain $\mathcal{D}$ are obtained by minimizing the slope with the boundary condition $\tilde{H}(\partial \mathcal{D}) = H(\partial \mathcal{D})$, where $\partial \mathcal{D}$ denotes the boundary of the domain $\mathcal{D}$. Assuming a smooth decay without saddle points below the shore line, a reasonable approximation of the topography in the area $\mathcal{D}$ is given by the Laplace equation

$$\nabla^2 \tilde{H} = 0 \qquad \tilde{H}(\partial \mathcal{D}) = H(\partial \mathcal{D}). \tag{4.17}$$

Apart from a saddle point free interpolation which satisfies the boundary conditions, the solution of equation (4.17) minimizes sharp edges in the topography leading to a smooth transition between the fixed boundaries.

For solving the equation numerically we used a successive over-relaxation scheme (SOR) directly on the numerical data grid, provided by the DEM to solve (4.17) for $\tilde{H}$. Convergence was defined for a total residual of less than 10^{-15}. The volume of the lobe deposit is then computed as the difference between the DEM of the Lobe surface and the interpolated original surface which yields a total volume

$$V_D = 8.7 - 9.3 \times 10^{10} \, m^3. \tag{4.18}$$

Good agreement between the two methods confirmed that the assumed long-term sediment supply $\overline{Q_s}$ was a good estimate for the sediment input during the Balize Lobe formation.

4.4.2 Comparison of variables and parameters with the Balize Lobe

To rescale the variables and parameters of the model, we ran a simulation of the model until a birdfoot delta was formed which compared well visually with the Balize Lobe of the Mississippi. The unitless horizontal resolution of the model was then isotropically rescaled by comparing the lattice spacing of the simulation grid with the real extent of the Balize Lobe. The simulation domain corresponded to a square segment of the Mississippi Delta of about $110\,km \times 110\,km$ with an error of about $10km$, with the river running along the diagonal of the grid. The horizontal rescaled grid resolution is $r = r_x = r_y = 380 \pm 10m$.

The water current at the top of the delta I_0 was rescaled to match the long-term mean annual discharge $\overline{Q}$ with the scaling constant $c_I = 1 \times 10^8\,m^3/s$, and then applied to all water fluxes I_{ij}. Note that the rescaling of the water flux is only necessary to obtain the correct units, as the model assumes a steady state for the water flow in each time step. Similarly, the sediment input at the top of the delta s_0 was rescaled to match the long-term mean annual sediment load $\overline{Q_s}$ with the scaling constant $c_s = 1 \times 10^4\,m^3/s$, before being applied to all sediment fluxes J_{ij}. The rescaled variables and parameters are summarized in Table 4.1 at the end of the chapter. Parameters are presended as ranges due to uncertainties in the data.

The rescaling factor in the elevation was obtained by comparing the highest and deepest points of the simulation domain with the maximal values in the DEM model. The slope of the simulation grid is given by the elevation difference between the upper left H_a and the lower right H_b corner divided by the length of the diagonal. Using the distance between the highest and lowest points of the simulation domain, $d = 150 \pm 5\,km$, $H_a = 0.058$ and $H_b = -0.1$, we obtain a slope of $s = 0.8\text{-}1.4 \times 10^{-6}$. To obtain a comparable slope from the DEM model, we used the highest and lowest point in the chosen segment of the map outside of the deltaic deposits which are $z_{max} = +5\,m$ (point (a) Fig.4.15) close to New Orleans and $z_{min} = -60 \pm 10\,m$ (point (b) in Fig.4.15) on the sea side at the downstream end of the Balize Lobe. This yields an average slope of the Mississippi $s_{Miss} = 3.3\text{-}5.3 \times 10^{-4}$.

4.4. INTERPRETATION OF THE MODEL PARAMETERS

Since both slopes should be equal, all elevations within the simulation have to be rescaled by a factor of $c_h = 380 \pm 10\,m$. As a consistency check, the rescaled lengths obtained by aerial pattern comparison can now be used to compare the amount of subaqueous deposits in the delta region. The topography of the simulation data which corresponds to today's Balize Lobe is now subtracted from the initial configuration to determine the volume of the simulated deltaic deposit. Using the rescaled variables one obtains a total deposited volume of

$$V_{\text{sim1}} = 9.6 - 10.2 \times 10^{10}\,m^3 \qquad (4.19)$$

which is comparable to the values obtained from the Mississippi measurements. The rescaling factor for the time step in the model c_t comes from the age of the lobe τ. In the simulation a delta stage similar to today's Balize Lobe is obtained after 27 million steps, which then yields a time resolution scaling constant $c_t = 1000 \pm 50\,s$. From the modeled sediment flux we can estimate the sediment supply to the delta by integrating the fluxes over the age of the Balize Lobe using the scaling factors obtained for the vertical, horizontal and temporal scales. This yields a supplied volume of 34-40×$10^{10}\,m^3$. Following the measurements of Corbett [40], 40-50% of the incoming sediment is advected beyond the delta lobe. Thus the simulated delta volume is in the range

$$V_{\text{sim1}} = 11 - 16 \times 10^{10}\,m^3. \qquad (4.20)$$

This is marginally higher than the volume obtained by comparing the surfaces which means that in the simulation more sediment is lost through the boundaries than the assumed 40-50%.

4.4.3 The erosion-sedimentation process

The erosion-sedimentation rule in Eq.(4.6) is the fundamental equation in the model that drives delta formation. It was motivated by the approach of Foster and Meyer [66] who proposed that the erosion-deposition rate should be proportional to the difference between the sediment transport capacity and the actual sediment load.

The histogram of the flow rates in Fig.4.19 shows that both erosion and deposition occur simultaneously in the simulated delta. Erosion will be the dominant process in the existing channels with a high flow rate, while deposition occurs over large areas of the forming lobe where $I_{ij} < I^*$ and

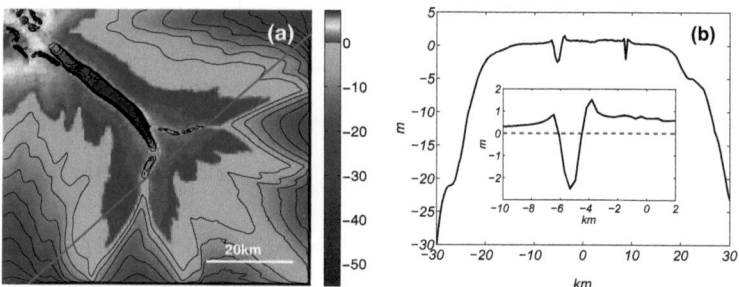

Fig. 4.16: *Topography of the simulated delta lobe (a) after 800 years. The contour lines of the subaqueous deposits are marked with black. A cut through the delta lobe along the red line is shown in (b). The inset zooms into the area close to the south flowing channels where the levees can clearly be identified.*

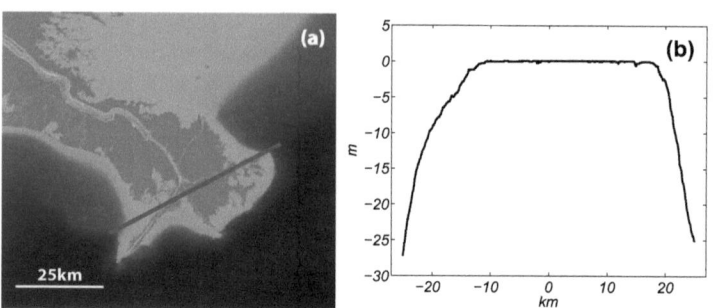

Fig. 4.17: *Figure of the Baliye Lobe of the Mississippi Delta (a) and cross section through the Balize Lobe (b). The cut is taken along the red line in map (a).*

4.4. INTERPRETATION OF THE MODEL PARAMETERS   55

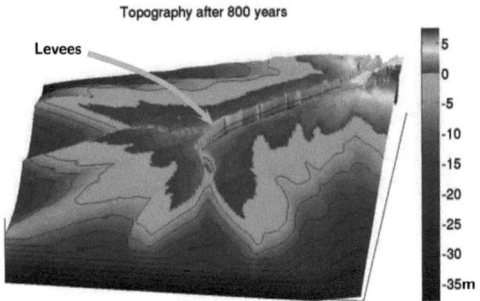

Fig. 4.18: *Three dimensional view of the deposited delta lobe, where the subaerial and subaqueous levee formation is clearly visible.*

$I^* = 4 \times 10^{-6}$ ($I^{*'} = 400\,m^3/s$). The parameter I^* is crucial for tuning the balance between these two processes. From Fig.4.19, it is noteworthy that the histograms of the water I_{ij} and the sediment J_{ij} fluxes follow a power law over a wide range of scales. This is indicative of the complex structure of the developing delta. The probability distribution functions (pdf) were obtained by integrating over time and space over the whole delta formation period. A similar behavior with fat tailed sediment flux distributions has been obtained [81] for avulsion of braided rivers.

Finally the distribution of dS^*_{ij} is shown in Fig.4.20. Over the entire simulation domain the frequency of the deposition was $P(dS^* > 0) = 0.0533$, while that for erosion was $P(dS^* < 0) = 0.0071$. This confirmed that deposition was the dominant process in the model. $P(dS^* = 0) = 0.9396$ are cells where no erosion or deposition is taking place, neither on land nor in the stable parts of the newly formed delta. The peak on the deposition side in Fig.4.20 corresponds to the deposition capacity $dS_{\max} = cI^*$ in Eq.(4.7). Note that the infrequent high rates of erosion $dS < \theta$ are not actually applied in the model.

When comparing the simulated water and sediment fluxes with observations it is important to recall that the model cannot resolve subgrid variability. The distribution channels in the Mississippi Delta are often narrower than the grid resolution $\Delta x'$ and are therefore more likely to have higher flow velocities and sediment transport [124].

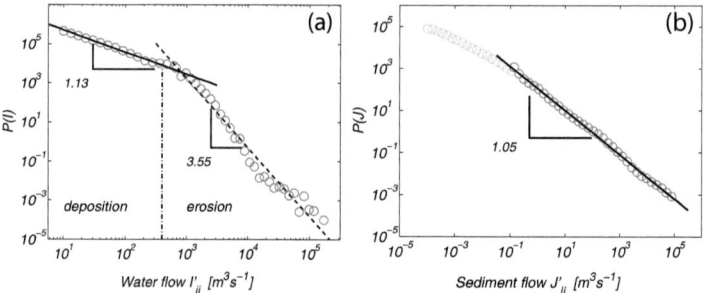

Fig. 4.19: Histogram of the magnitude of the water (a) and sediment flow rates (b) in the deltaic plain integrated in space and time over the whole delta formation process. While the water shows a power law decay close to one (solid line) for small flow rates the exponent for very large flows is 3.55 (dashed line) for over 2 decades. The dotted vertical line corresponds to the flow rate I^*. The distribution of sediment flows shows a power law tail with exponent $\gamma = 1.0$ over 4 orders of magnitude. The solid black line shows the least square fit of the tail (dark gray circles) to data.

4.4. INTERPRETATION OF THE MODEL PARAMETERS

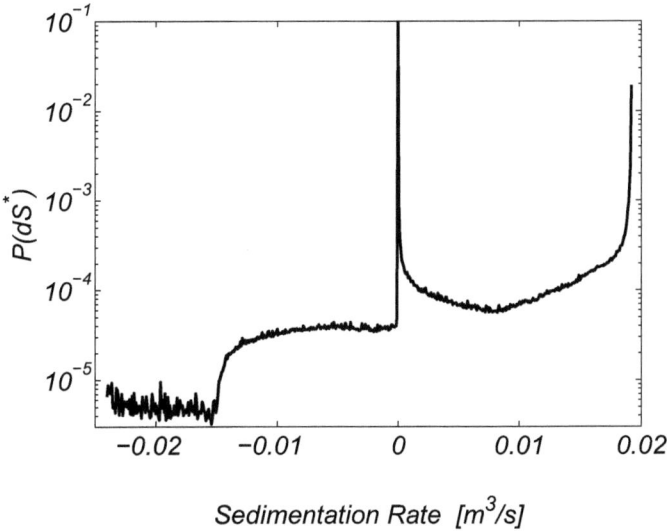

Fig. 4.20: *Relative frequency of the magnitude of the sedimentation- erosion rate dS^* averaged over time for the whole delta region. The maximal deposition is given by the product of the threshold I^* and the sedimentation constant c $dS_{max} = cI^*$. The strong peak at zero corresponds to the nodes with no flow (dry land).*

Table 4.1: Rescaling of the dimensionless parameters in the model for the Mississippi birdfoot delta.

	Model	Rescaled variable	Scaling constant
Horizontal scale	$\Delta x = 1\,gp$	$\Delta x' = r \cdot \Delta x$	$r = 370 - 390\,m/gp$
	$\Delta y = 1\,gp$	$\Delta y' = r \cdot \Delta y$	
Vertical scale	H_i	$H'_i = c_h \cdot H_i$	$c_h = 370 - 390\,m$
	V_i	$V'_i = c_h \cdot V_i$	
Time scale	$dt = 1$	$dt' = c_t \cdot dt$	$c_t = 950 - 1050\,s$
Water flux	$I_0 = 1.7 \times 10^{-4}$	$I'_0 = c_I \cdot I_0$	$c_i = 1 \times 10^8\,m^3/s$
	I_{ij}	$I'_{ij} = c_I \cdot I_{ij}$	
	$I^\star$	$I^{\star\prime} = c_I \cdot I^\star$	
Sediment flux	$s_0 = 2.5 \times 10^{-4}$	$s'_0 = c_s \cdot s_0$	$c_s = 1 \times 10^4\,m^3/s$
	J_{ij}	$J'_{ij} = c_s \cdot J_{ij}$	
Erosion-deposition rate	dS_{ij}	$dS'_{ij} = ed S_{ij}$	$\epsilon = \dfrac{r^2 c_h}{c_t} = 4.8 - 6.6 \times 10^4\,m^3/s$
Erosion-deposition parameter	$c = 0.1$	$c' = \dfrac{\epsilon}{c_i} c$	$c' = 4.8 - 6.6 \times 10^{-5}$

4.5. LOBE SWITCHING AND LONG-TERM TEMPORAL DYNAMICS

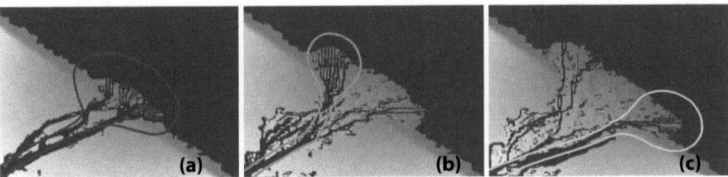

Fig. 4.21: *The figures show the switching of the delta lobe during the simulation. Comparing Figs. (a) and (b) a type I switching can be identified where the main part of the delta lobe is abandoned close to the mouth of the river just before the river splits into several distributaries and forms a new lobe beside. Another type of delta switching is shown in Figs. (b) and (c). Here the channel switches far upstream and takes a new course to the coast forming another delta lobe far away. This is referred to as a switching of type II. The parameters for the simulation where $I_0 = 1.7 \times 10^{-4}$, $s_0 = 5 \times 10^{-5}$, $c_2 = 0.0005$, $c_1 = 0.0005$, $c_1 = 0$ and $I^* = 3.3 \times 10^{-4}$. The simulation was run on a 179×179 lattice with smoothing every 2000 time steps and a smoothing constant of $\epsilon = 0.0001$.*

4.5 Lobe Switching and Long-term temporal Dynamics

The rich dynamics of the switching phenomenon observed in the Mississippi were also identified in our simulations. Figures 4.21(a-c) show some snapshots of a delta simulation, where the different types of lobe switching can be identified. A lobe shift of type I for example can be found in Figs.4.21(a& b), while between Figs.4.21(b&c) the stream switches far upstream related to a type II lobe shift.

A type III lobe switch occurs in the simulation of the birdfoot delta (Figs4.23(d&e)) where the channel to the east was abandoned and the sediment is transported exclusively through the south distributary. Small crevassing appears during the whole simulation at different places. A typical snapshot is shown in Fig.4.23(f) where the river breaks through the channel bar and fills the adjacent bay.

The delta formation process by erosion and deposition can be divided into several stages where each stage is dominated by different effects. The early stage of the delta formation process is dominated by subaqueous deposition of sediment along the coast and the accumulation of a large amount of sediment on the continental shelf. Due to the high sediment supply by the river plume (Figs.4.23(a & d)) deposition also takes place far away from the

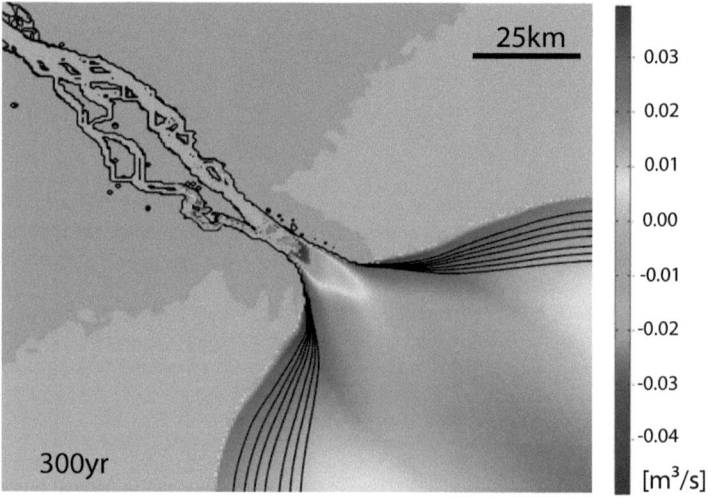

Fig. 4.22: *The figure combines the two main aspects of the sedimentation and erosion process in the delta, which are (a) the sediment supply by the river, given by the contour levels (black) of the sediment flow and (b) the erosion-sedimentation rate determined by the river current. The actual applied deposition rate is marked with the color range. The subaqueous deposition and the growth of the prodelta far beyond the actual mouth of the delta is clearly visible. Deltaic deposits form a complex subaqueous landscape with levees and bars confining the flow in the channel bed (c.f. Fig.4.18).*

river mouth leading to the accumulation of thick prodeltaic deposits which form the base of the future delta. Fig. 4.22 shows the spatial distribution of the erosion-deposition process at the delta mouth after 300 years of delta progradation. Subaqueous mouth bars form in the model when the flow diverges after the birdfoot lobe has prograded a certain distance into the sea, but they do not always become apparent above the water level.

When the stream enters into deep ambient water the current is immediately decelerated leading to high deposition rates and fast subaqueous delta growth along the coast. In our simulation, this effect can be observed in the early delta formation stages (Figs.4.23a and b) when the river enters into the sea where the conductivity σ_{ij} is much higher and the flow is distributed over a wider area. This leads to a decrease in the local water flow I_{ij} resulting

4.5. LOBE SWITCHING AND LONG-TERM TEMPORAL DYNAMICS

in an increasing deposition rate. As there is still a slope in the direction of the main flow, the lateral deposition rate is slightly higher forming subaqueous levees that confine the flow in the main channel direction. In general the morphodynamic feedback of the inflowing current leads to the formation of subaqueous landforms that confine the initial flow.

The next phase is dominated by deposition in the channels and subaqueous sedimentation that leads to a more gradual transition from the river into the ocean. Channel deposition also limits the maximal distance a river channel segment can prograde before its slope becomes too shallow to transport sufficient sediment for further lobe growth. Strong deposition at the end of the channel head leads to a splitting of the stream, and subsequent overbank avulsion that causes lateral growth of the lobe. For the formation of a birdfoot delta it is important how the bank deposits force the channels to maintain their particular course by depositing levees along the channel. The formation of natural levees is inherent in our model as can be seen in Fig.4.16. As the lateral current in the model is much slower than the current in the direction of the main slope these flows lead to the deposition of levees confining the flow in the main direction.

When the stream approaches the shoreline and deposits more sediment on the channel bed this sedimentation eventually overtops the levees. Bank overflow, rapid bank and channel bed aggradation, ultimately induce the failure of the natural levee and a new outflow channel for the stream forms with a steeper slope. This phenomenon was observed several times during the simulation (Fig.4.23(e&f)). In the literature this shift in the main sediment transporting channel is often referred to as Type III lobe switching [32]. A cut through the delta lobe after 27 million simulations steps, corresponding to about 800-1000 years, is shown in Fig.4.16. The subaerial part is flat and the channels are confined between natural levees Fig.4.16(b), then the lobe drops steeply. The cut is indicated with a red line in Fig.4.16(a). For comparison, a cut through the Balize lobe of the Mississippi created from the bathymetry data [53] is shown in Fig.4.17. The profile shows a morphology with steep drop-offs and a flat delta surface, similar to that in the simulation.

The different phases of the delta cycle can be identified in the time series of the land growth illustrated in Fig.4.24. The change of the land/sea fraction is plotted versus time. Formally the growth rate is defined as follows: Let N be all nodes on the lattice, then the land/water fraction at a time t_k is defined

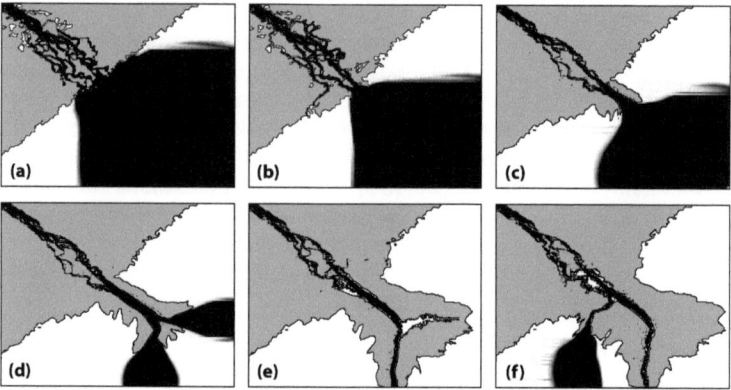

Fig. 4.23: *The time evolution of a birdfoot delta (from left to right and top to bottom). In (a) the coastline of the delta after 50 years is drawn, and shows the initial stage of the delta formation. Between 50 years and 100 years (b) the delta mainly deposits along the coast while after 300 years(c) the main channel progrades into the sea depositing sediment mainly on its levee sides. After 700 years the main channel splits into two distributaries (d), whith the smaller one becoming inactive after 1200 years (e). A new channel breaks through the sidewalls after 1500 years (f). The area of active sediment transport is marked in black. Here one can see how sediment flux emerges in new channels that are abandoned later.*

4.5. LOBE SWITCHING AND LONG-TERM TEMPORAL DYNAMICS

by:

$$W(t_k) = \frac{\sum_{n=1}^{N} \Theta(H_i - V_i)}{N} \quad \text{where } \Theta(x) = \begin{cases} 0 & \text{if } x < 0 \\ 1 & \text{if } x \geq 0 \end{cases} \quad (4.21)$$

where N is the total number of all nodes in the lattice. Thus the growth rate is given by the difference $W(t_k) - W(t_{k-1})$. After a phase of constant growth, the stream breaks through the river banks and searches a new path, accompanied by high deposition and the formation of new subaerial delta parts corresponding to the peaks in Fig.4.24. When the shallow water behind the levee is filled up and the stream enters into deeper water the surface growth of the delta decreases tremendously. The delta starts to build up a new subaqueous deposit until the lobe enters into a phase of constant lobe growth. Since the land formation process occurs in bursts, we investigate the possibility of long-range ctemporal orrelation with the delta growth. This can be characterized by the Hurst exponent, H, which was introduced by Hurst to describe the water level fluctuations of the Nile [78].

The method used here to detect and quantify the correlation properties in a non-stationary time series is detrended fluctuation analysis (DFA) [125; 77; 31]. It is based on the computation of a scaling exponent H', which is equivalent to the Hurst exponent, by means of a modified mean square analysis. This method avoids spurious long-range correlations due to non-stationary ties, by first integrating the data of the time series

$$y(k) = \sum_{i=1}^{k} [x(i) - M] \quad (4.22)$$

where

$$M = \frac{1}{k} \sum_{i=1}^{k} x(i) \quad (4.23)$$

is the average value of the series $x(i)$ and k ranges between 1 and N. Next, the time series is mapped onto a self-affine stochastic process by dividing the integrated time series into K equally spaced intervals of length n. In each of these boxes the least square fit (linear, quadratic, etc.) is performed on the data and the y-coordinate of the resulting curve segments in box k is then called $y_n(t)$ where n denotes the length of the segment. In order to detrend the time series $y(t)$, the local trend $y_n(t)$ is subtracted in each of these boxes. The root mean square (RMS) fluctuation is then calculated as,

$$F(n) = \sqrt{\frac{1}{N}\sum_{t=1}^{N}(y(t) - y_n(t))^2}. \tag{4.24}$$

Scaling is present if $F(n)$ has a power law dependence on the size of the time window n,

$$F(n) \propto n^{H'} \text{ for } n \to \infty, \tag{4.25}$$

where $H' \approx H$ asymptotically. The exponent H is related to the "$1/f$" noise spectral slope D_f (fractal dimension) by

$$D_f = 2H - 1. \tag{4.26}$$

It can be shown that in case of no correlation, such as, for example, in a pure random walk, the exponent H is exactly 0.5. The exponent for a positively auto-correlated time series is H >0.5. This means that if there is a trend to increase/decrease from time step t_{i-1} to t_i there is a higher probability to follow the trend than in a completely random process. A Hurst exponent of H <0.5 will exist for a time series with anti-persistent behavior (negative autocorrelation), which means that an increasing trend will be followed more probably by a decrease and vice versa. This behavior is sometimes called "mean reversion". To improve the statistics for the scaling exponent we ran five different realizations of the birdfoot delta formation with the same parameters but different seeds for the random numbers of the initial surface. For each sample the DFA analysis was performed and the averaged result is plotted in Fig.4.24. Two different correlation regimes can clearly be distinguished. On short time scales the land formation process is highly correlated with exponent H =1.2 which corresponds to the formation of a new outflow channel and a strong land growth. On longer time frames the different cycles average out and yield a smaller exponent which was determined to be H =0.7 in the simulation of a birdfoot delta. It seems that the stronger long-range correlations for shorter time scales is the result of gradual lobe building phases at the mouths of the distributary channels, while at some temporal scale the random avulsion mechanisms lead to a breakdown of the correlation structure. As the delta grows in bursts of crevassing and avulsion it is interesting to analyze the waiting time distribution between such events and their spatial and temporal correlations in more detail. This work is still in progress.

4.5. LOBE SWITCHING AND LONG-TERM TEMPORAL DYNAMICS

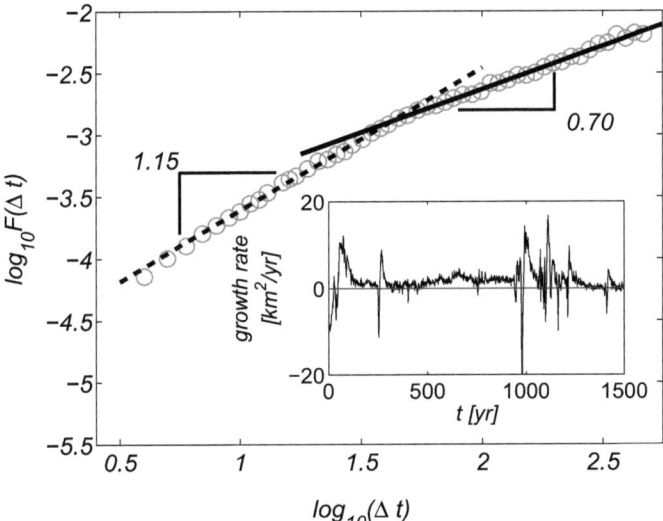

Fig. 4.24: *The inset shows the signal of the delta growth rate versus time while the main plot is the corresponding DFA analysis averaged over 5 samples. The error bars of the sampling are not shown in the plot as they are smaller than the plot symbols. The straight line is the power law fits of $f \sim \Delta t^\alpha$ to the data with exponents $\alpha = 1.15$ (solid line) and $\alpha = 0.70$ (dashed line) The plot shows clearly two regimes with different correlation exponent. For small time frames delta growth is highly correlated while on longer time scales the effects of the deltaic cycles average out the strong effects.*

Chapter 5

Inland Delta generation with application to the Okavango

Some deltas form inland. The Okavango River of Africa for example flows into a landlocked basin, where it soaks into the ground or evaporates. The word 'delta' is used for this type of deposit because the aerial pattern resembles a river delta, but they are correctly classified as alluvial fans. Alluvial fans occur in areas where topographical slopes are extremely flat as a result of high local subsidence.

The Okavango Delta is one of the world's great wildlife habitats. However with greater industrial development in Africa the demand for electrical energy and water consumption are on increasing. Construction of dams and extraction of water may have tremendous influence on the wildlife in the fan area. Therefore, a detailed study of the processes and products and the interplay between thr two are not only important for the assessment of water management but also for the preservation of wildlife habitats.

We extended the application of the delta simulation model to describe the formation of inland deltas. Evaporation and seepage have been added to the flow equations and the erosion laws have been adapted to describe the drying up of old channels and creation of new ones.

5.1 The Okavango Delta

The Okavango Delta is one of the world's largest inland deltas[71]. Its headwaters start in Angola's western highlands, with numerous tributaries joining to form the Cubango River, which then flows through Namibia, where it is called the Kavango. Finally it enters Botswana, where it is then called

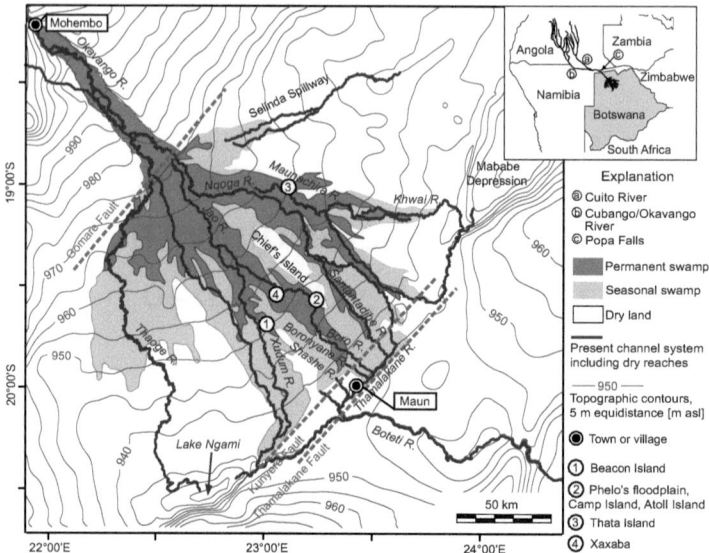

Fig. 5.1: *Map of the Okavango Delta. Green: channels and swamps; blue lines: permanent channels; dashed gray lines: tectonic faults. Figure after [108]*

the Okavango. Millions of years ago the Okavango River flowed into a large inland lake called Lake Makgadikgadi, today Makgadikgadi Pans. Tectonic activity and faulting interrupted the flow of the river causing it to backup and form what is now the Okavango Delta. This has created a unique system of water ways now supporting a vast array of animal and plant life in a region which would otherwise be a dry Kalahari savanna.

5.1.1 Hydrology of the Okavango

The Delta's floods are fed by Angolan rains, which start in October and end in April. The floods cross the border between Botswana and Namibia in April and will only reach the distal part of the Delta at Maun around July, taking almost three months to travel from the top of the fan to the bottom Fig. 5.1. The slow pace of the flood is due to the gradual gradient of the terrain, which drops little more than 60 meters over a distance of 450 kilometers. If there

5.1. THE OKAVANGO DELTA

is any outflow beyond Maun, it vanishes into the Kalahari along the course of the Boteti River. The geomorphological evolution of the surface is mainly influenced by river dynamics and their transport capability, where local incline and the material properties of the soil play an essential role. Sediment is brought to the Delta mainly during the flood peak each April. Inside the Delta, channel discharges show low annual variation because water overboarding water is stored in floodplains. This overland flow as well as the water transport underneath floating papyrus has been found to be significant for bedload and suspended sediment transport [107]. The total sediment transport capacity of the relevant channels decreases downstream which accounts for the accretion of the Delta. The sediment of the Okavango Delta consists of fine-grained sand with a medium diameter of $0.35mm$. River branches display large variation in channel width and roughness even over short distances.Channel width varies between $10m$ and $100m$. There is also large seasonal variation in the hydraulic heads of the Kalahari Sand aquifer close to the channels whereas at the fingers the hydraulic heads remain relatively constant throughout the year around $20m$ below the surface [108].

With a total size of more than $40,000km^2$ the alluvial fan located at the end of the Okavango River is one of Africa's largest inland water and represents an unusual depositional setting in which peat-forming perennial swamps of around $6000km^2$ occur in a region of aeolian and semiarid sedimentation. The whole system is located within an incipient graben or rift valley, which in turn is a part of the southern continuation of the East African Rift. The Delta itself is arguably the most pristine of Africa's large wetlands, with an naturally unregulated hydrological regime. The malaria mosquito and Tsetse fly which are prevalent throughout the Delta historically discouraged human settlement [102]. It is remote from regional population centers and did not attract colonial settlers. Important economic activities in the Okavango region are fishing, sport hunting and tourism. The Okavango River drains a relatively large seasonal inflow of about $10.5 \times 10^9 m^3/yr$ [52] of water into the wetlands [159; 101], though 96% of this is lost where 2/3 of the loss is caused by evaporation [159] and 1/3 due to transpiration [108]. A further 2% is lost by water seepage [159].

The vegetation in the Okavango swamps consisting of various aquatic species [139], affects the hydrology of the Okavango because most of the water is lost due to evapotranspiration by the plants. Bauer et al. [16; 14; 15] estimate the local evaporation rates from NOAA remote sensing images.

Fig. 5.2: *Aerial photo of the Okavango Delta. Photo courtesy of PeterSp.*

5.1.2 Geology of the Okavango

Africa is currently being split apart by geological processes along the Great Rift Valley starting north in Ethiopia, and passing through Kenya, Uganda and Tanzania. In Tanzania, the Great Rift Valley divides and cuts through Malawi and Mozambique to the East and to the Southwest through Zambia's Luangwa Valley and Kariba Valley to the Victoria Falls. From Victoria Falls, it continues southwest to form the rift-system through the Chobe-Linyanti area to the Okavango Delta at its southwest extremity where the Great Rift Valley reaches its lowest elevattion. During the last 3 million years, the Delta has formed at the end of the rift where it cuts through the center of the Kalahari sand sheet in a southwest direction. The main Okavango rift lies between two parallel fault lines – one, the Thamalakane fault, at the lower extremity of the Delta passing through Maun and the other, the Gomare fault, at the northern edge of the Delta along the line of the Selinda Spillway. The area between the two southwesterly faults lies about 300 meters below the plain.

Geological tension splitting the southern African continent in the East-West direction rather than NW-SW like the rift. This produces a series of conjugate faults perpendicular to the Okavango rift. The most prominent of these faults have created a minor rift forming the Okavango "panhandle" or the

5.1. THE OKAVANGO DELTA

main Okavango River which empties into the Delta from the Northwest. The area immediately east of this rift became slightly uplifted forming Chief's Island.

Before the formation of the Okavango Delta, the river was flowing directly through the area where the Delta is now, passing the Lakgadikgadi Pans and continuing out to the Indian Ocean via the Limpopo River. When the Thamalakane fault at the lower edge of today's Delta rose up, the Okavango River was blocked, initiating the Delta's formation.

As the Okavango's rift subsided, it filled up with a mix of windblown sand from the Kalahari and sediments carried by the waters of the Okavango River until it attained its present form as a major inland alluvial fan. The major parts of the Delta today are flat, dry Kalahari sand sheets intersected by a shallow rift valley trough with a major river flowing into it along the Panhandle fault line. Water carried by the river gives rise to the biological productivity and beauty of today's Okavango Delta. About 30,000 years ago, Northern Botswana was much wetter than it is today and the Delta wetlands were much larger. A huge inland lake extended from the Kwando / Linyanti Rivers via the Savuti Channel to the Mababe Depression along the Thamalakane River to Lake Ngami and via the Boteti River to greater Lake Makgadikgadi.

After a number of climate fluctuations, the last wet period was around 2,000 years ago when the Mababe Depression formed a lake connecting to Lake Ngami via the Thamalakane River. Since then the Delta has been progressively drying up.

5.1.3 Sedimentation processes in the Okavango Delta

All sediment carried by the Okavango river is deposited on the fan. Sediment distribution and the dispersal of sediment in the fan was investigated by McCarthy[101]. Overlying the bedrock the base of the fan is formed by a thick layer of medium and fine-grained sands and silts of Cenozoic age collectively known as the Kalahari bed [79]. Using a 3D Euler Deconvolution Technique, Modisi [109] and Atekawa [10] estimated the thickness of the Kalahari bed to be around $300m$ on average along the graben. Kgothang [84] estimated an average $150m$, applying the same technique on the entire Delta area.

At present over 870,000 tonnes of sediment are deposited every year in the Okavango Delta where the bedload exceeds the suspended load by at least a factor of four. Half of the total mass delivered to the Delta arrives in dissolved state including silica, calcium and magnesium bicarbonates[101]. The total bedload is estimated at $200,000 t/yr$ and the solute load exceed

450,000t/yr. At 50,000t/yr the suspended load is almost negligible. The remaining 170,000t/yr are aeolian wind blown sediments imported from the Kalahari desert into the Delta. The flow velocities in the channels range from 0.5m/s to 1.5m/s and are of the order of a few centimeters per second in the floodplains. As a result, bedload is uniquely transported in the channels. The small fraction of suspended load is filtered through permeable levees stabilized by vegetation and peat confining the channels to their beds. Solutes are deposited as calcites or silcretes mainly in the floodplains and beneath islands where water loss due to evapotranspiration enhances the precipitation in the soil. Sodium bicarbonate accumulates in the ground water beneath the island centers, which impacts on the vegetation, leading ultimately to barren island interiors[102].

In contrast to other alluvial fans in semi-arid climates, the Okavango fan is extremely flat. The elevation drop is less than 60m from the entrance close to Mohembo at 1,000m to 940m at the outerfringe of the Delta at Maun. The Okavango Delta swamps are divided into three parts [158]: The upper Panhandle, where the swamps are confined between the shoulders of Kalahari sand, which rises about 4m above the floodplain and represents fault scarps [79]. Here, the Okavango Rsiver takes a meandering course. The Panhandle is characterized by wide channels between 50m and 100m, typically 3-4m deep with a maximum flow velocity between 0.6m/s and 0.8m/s. Channels have sandy beds and are flanked by dense stands of *Cyperus papyrus* rooted in mud-rich peat. The floodplain of the Panhandle is flanked by banks of semi consolidated Kalahari sediment. At the southern end of the Panhandle, the swamp broadens as it crosses the Gomare Fault forming the alluvial fan [103]. The second part consists of largely unconfined perennial swamps and, finally, the unconfined seasonal swamps of the lower floodplain. The average gradient across the Delta is about 1:3500 [159]. Here permanent swamps develop in the proximal area, with branches extending along the major distributary channels. The northeastern region of the permanent swamps is supplied by the Nqoga and Maunachira channels, typically 15-50m width and 2-4m deep. These channels have sandy beds and the margins are densely vegetated by Papyrus or Miscanthus, forming vegetation levees [100; 141]. Backswamp areas behind these levees tend to be more open and are colonized by a variety of aquatic species [60]. Large lakes exist within this permanent swamp representing ancient oxbows [99]. Overspill from this region of the swamp supplies the Maunachira- Santantidibe system to the east of Chief's Island. The swamps in the central region of the Delta show a comparatively poorly developed channel system, the Jao-Boro

5.2. MODELING INLAND DELTA FORMATION

channel, and the channel widths are generally less than 15m. Extensively flooded areas called Xo flats, characterize this region.

The sediment is mainly distributed through main channels from which water and sediment are deposited in splashes into the swamps during high floodstages and breaches in the levees. Fine-grained sand and clay is deposited in different distributaries raising the channel bed until the channel ceases to transport sufficient water and sediment, dries up and switches to a new path through the Delta. This type of switching is similar to the Type III lobe switching of marine deltas: the main course of the sediment flow gradually shifts over the entire fan. Uplift and subsidence of parts of the Rift Valley structure diverts the distribution of water within the different channels over a longer time.

5.2 Modeling Inland Delta Formation

5.2.1 Modification of the River Delta Model

For simulating inland delta formation the presented model was modified to include evaporation and seepage. In each node, a source/sink term E_{ij} is applied

$$E_{ij} + \sum I_{ij} = 0. \tag{5.1}$$

In the relaxation form of Eq. 4.4, Eq. 5.1 reads as follows

$$V_i' = V_i + \delta \sum I_{ij} + E_{ij} \tag{5.2}$$

Note that in the case of evaporation E_{ij} is negative. The strength of the sink can be variable in each node depending on flow conditions and water depth. The fact that the channels now end in the desert results in new difficulties at the end points of the distributary channels. These end points of the distributaries are called "dead ends". When coupling the flow with the erosion law two different types of dead ends have to be distinguished: invasive and decreasing inflow. In the first type, the influx into the end node increases from one sedimentation step to the other. In this case there is no deposition in the final node whereas in the second case of decreasing inflow the channel dries up and deposits its load also at the end point.

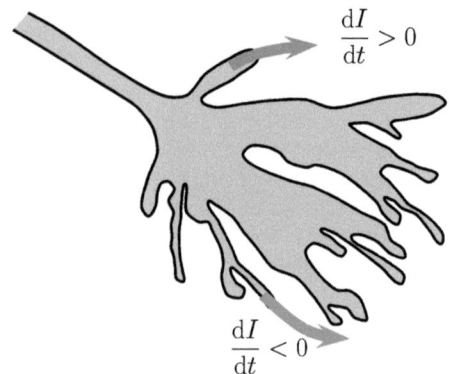

Fig. 5.3: *Sketch of the Okavango Delta. Two different channel ends have to be distinguished. Sedimentation only takes place in channel ends which dry up e.g. $dI/dt < 0$. In the case $dI/dt > 0$ no sedimentation occurs at the end point.*

5.2.2 Simulating Inland Delta Formation

The system was initialized with an inclined plane with the main slope running along the diagonal of the lattice. The ground water table was set to be $\varepsilon = 0.0025$ below the surface and the elevation increased slightly sidewards, perpendicular to the main slope as in the simulation of the coastal river deltas. On the boundary the water level was kept constant below the ground, so water will simply seep away when reaching the boundary margin. A constant evaporation rate E_{ij} was applied in each cell with surface water. We also implemented an evaporation rate that increases with its distance to the inlet node namely

$$E_{ij} = \frac{(i^2 + j^2)}{2N^2} \hat{E}_{ij} \tag{5.3}$$

where $\hat{E}_{ij}$ is the maximal evaporation and N the size of the lattice.

After the creation of an initial channel network the structure becomes almost stationary, shifting only slightly at the channel ends before depositing the load there. As a result of an extensive parameter study comparing the different combinations of water flux and evaporation rates coupled with different erosion strengths we found that a distinct channel network can only be obtained for erosive regimes, while in a deposition-dominated case the land is more equally flooded. Major shifts of the main channels over the

5.2. MODELING INLAND DELTA FORMATION

fan could not be observed in the simulation. This leads to the conclusion that the reduced hydrodynamic flow routing scheme of the model reproduces channel flow less accurately than topography-driven models based on the phenomenological Manning-Strickler type formulae used in models like CAESAR or EROS [44; 49; 59].

Figure 5.4 shows four examples of different water/land surface patterns for high/low erosion and high/low evaporation using Eq(5.3). The increasing evaporation rate does not significantly affect the channel pattern but keeps the flow more inside the domain.

We analyzed the channel structure by calculating the change of the cumulated channel width in a cross section of the domain with distance to the inlet. Figures 5.5 show the plots for three different time steps. First the cumulative channel width increases and the water is distributed from the inlet into the different channels. When the flow prograde further into the domain the total channel width decreases gradually due to the loss of water by evaporation and seepage.

The depositional pattern between 2 timesteps is shown in Fig.5.7, where the deposits of anastomosing (bottom right) channels can be clearly identified. Wheras the typical channel depth in the Okavango is less than a few meters and the average channel width is around 10-20m the simulated channels are almost as deep as wide. Nevertheless deposition of levees occurs along the simulated channels, which is also observed in the Okavango (Fig.5.6).

Statistical similarity between the simulated channel network and the Okavango is established through the fractal dimension calculated with the box counting technique. The plots of the scaling behavior of the model and the Okavango pattern are shown in Fig.5.8, where the number of cells N which are covered by water versus the cell size s yields a power law over more than 2 decades. A least square fit of a power law

$$N \sim s^{-D} \tag{5.4}$$

yields a fractal dimension of $D = 1.83 \pm 0.05$ for the Okavango data which is in good agreement with the simulation result of $D = 1.84 \pm 0.05$. The pattern of the flooded area of the Okavango was derived by a combined analysis of high resolution aerial photos from GoogleEarth™ and NOAA satellite measurements.

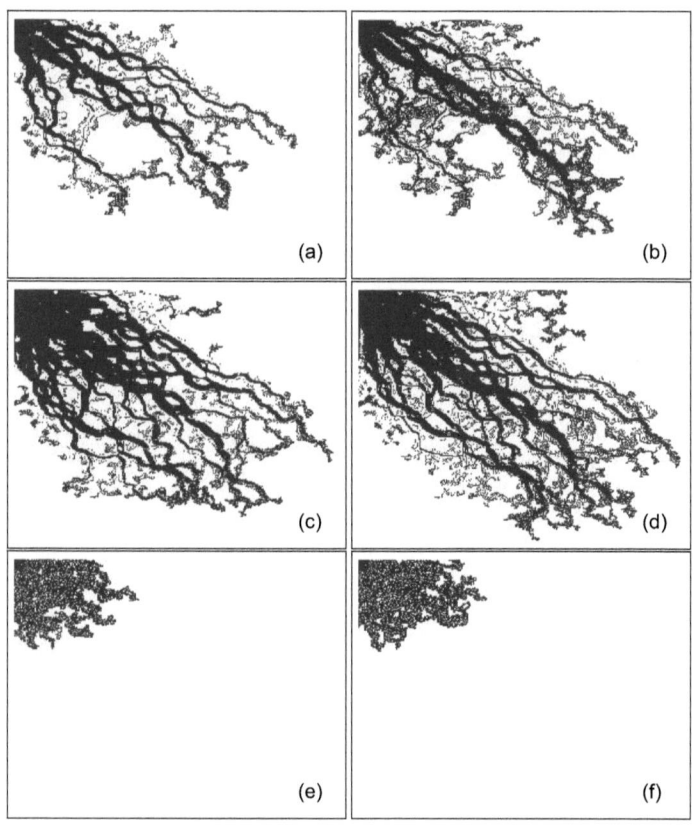

Fig. 5.4: *Simulation results for different parameter combinations (a) $I_0 = 1 \times 10^{-3}$, $I^\star = 5 \times 10^{-6}$, $Loss = 5 \times 10^{-8}$ (b) $I_0 = 1 \times 10^{-3}$, $I^\star = 7.5 \times 10^{-6}$, $Loss = 5 \times 10^{-8}$ (c) $I_0 = 2 \times 10^{-3}$, $I^\star = 5 \times 10^{-6}$, $Loss = 5 \times 10^{-8}$ (d) $I_0 = 2 \times 10^{-3}$, $I^\star = 7.5 \times 10^{-6}$, $Loss = 5 \times 10^{-8}$ (e) $I_0 = 1 \times 10^{-3}$, $I^\star = 5 \times 10^{-6}$, $Loss = 5 \times 10^{-6}$ (f) $I_0 = 1 \times 10^{-3}$, $I^\star = 7.5 \times 10^{-6}$, $Loss = 5 \times 10^{-6}$. The other parameters have been kept constant.*

5.2. MODELING INLAND DELTA FORMATION

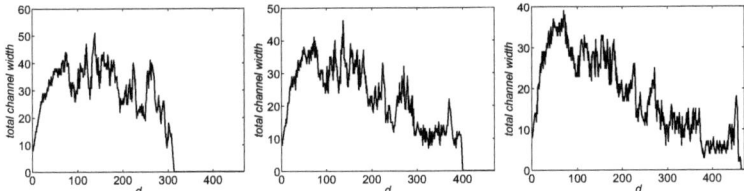

Fig. 5.5: *Plot of the cumulative channel width for three different time steps; (a) 5 Mio, (b) 7.5 Mio and (c) 10 Mio steps. After the inlet the flow is distributed through the inlet into the channels. The total channel width decreases gradually due to evaporation and seepage.*

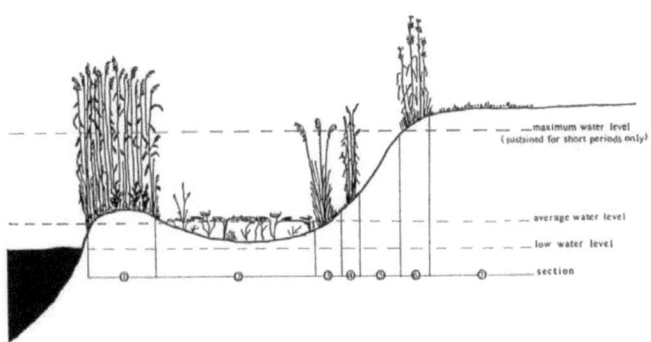

Fig. 5.6: *Cross-section of a typical channel in the Okavango. Levees are mainly formed by peat, covered with vegetation of Cyperus papyrus. The frequently flooded swamps are situated behind the levee. The figure is reproduced after [139]*

CHAPTER 5. INLAND DELTA GENERATION

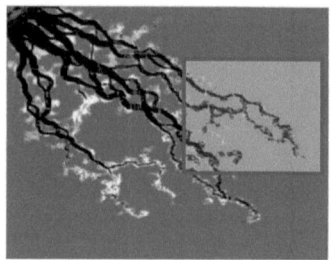 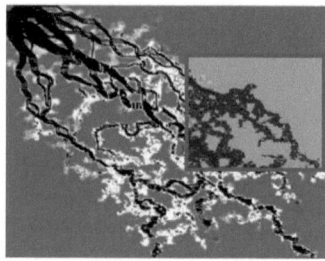

Fig. 5.7: *Change of the landscape in a simulation with evaporation after 10 million iterations (left) and after 50 million time steps (right). Blue indicates deposition, red erosion. The deposition at the channel ends and along the channels is clearly visible while erosion is dominant within the channels. Note the large deposition zone at the channel ends in the colored areas in the inset (right). Also the formation of levees along the channels can be observed along the streams.*

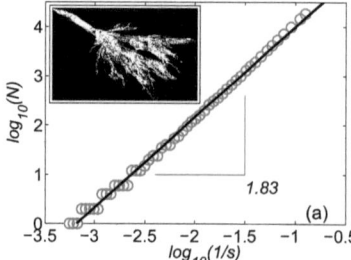 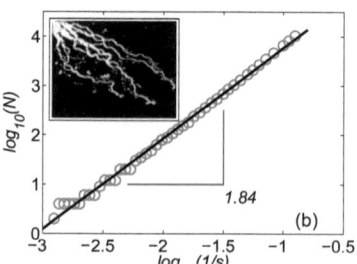

Fig. 5.8: *Fractal dimension of the land-water pattern in the Okavango (a) in comparison with the simulation (b)*

5.2. MODELING INLAND DELTA FORMATION

5.2.3 Rescaling of the variables

To gain further quantitative insight into the dry delta formation process we rescale the length and time scales as had been done for the Mississippi. The physical extent of the Okavango region is around 160×160 ± 20km including the panhandle or 130×130 ± 20km without. The Panhandle is a strictly confined region between two fault lines, which was not captured in the initial condition of the simulation. Therefore the simulation results were compared with the Okavango DEM data excluding the Panhandle. The simulation domain of 279 × 279 grid points yields a lattice spacing of $\Delta x \approx$ 600m. The elevation of the Okavango drops from 1000m at Mohembo to 940m close to Maun resulting in a mean slope of only s_{Oka} = 2.4 × 10^{-4}. When the Panhandle is excluded, a mean slope of 3 × 10^{-4} is obtained.

Using the constraint that the average slope of the model must be equal to the slope in the Okavango one obtains a vertical scaling factor of Δz = 600 ± 50m.

The inflow conditions of the model inlet were calibrated with the flows at the inlet of the panhandle close to Mohembo where the longterm average Okavango water discharge is 292 m^3/s. The sediment total load is estimated to be 870,000 tonnes per year[101]. The fluxes for water and sediment determine the rescaling constant for the system. The rescaling parameters are summarized in Table5.1. Using the rescaled variables, the simulation was compared with the real Okavango and a flume experiment in Chap.5.3.

Table 5.1: Rescaling of the dimensionless parameters in the model for the Okavango inland delta.

	Model	Rescaled variable	Scaling constant
Horizontal scale	$\Delta x = 1\,gp$	$\Delta x' = r \cdot \Delta x$	$r = 600 \pm 50\,m/gp$
	$\Delta y = 1\,gp$	$\Delta y' = r \cdot \Delta y$	
Vertical scale	H_i	$H'_i = c_h \cdot H_i$	$c_h = 600 \pm 50\,m$
	V_i	$V'_i = c_h \cdot V_i$	
Water flux	$I_0 = 5.0 \times 10^{-6}$	$I'_0 = c_I \cdot I_0$	$c_I = 1 \times 10^8\,m^3/s$
	I_{ij}	$I'_{ij} = c_I \cdot I_{ij}$	
Sediment flux	$s_0 = 2.5 \times 10^{-3}$	$s'_0 = c_s \cdot s_0$	$c_s = 1 \times 10^4\,m^3/s$
	J_{ij}	$J'_{ij} = c_s \cdot J_{ij}$	

5.3 Experimental Modeling

To complement the computational results a flume experiment was designed to including evaporation and seepage. Flume experiments on delta formation have been carried out in several institutions, as for example in the Earthscape Experiment in the St. Anthony Falls Laboratories and at Exxon Mobile where only recently a breakthrough in the experimental modeling of birdfoot deltas has been achieved. The group at Exxon Mobile found that the typical birdfoot structures can be reproduced in a flume experiment by increasing the cohesivity of the sediment [76]. Also recently a group in the Netherlands studied the formation of alluvial fans on Mars[90] but experimental work on inland deltas including evaporation had not previously been conducted.

Our flume consisted of a $1m$ by $1m$ basin fixed on a concrete base at an inclination of 6 degrees. The main slope ran along the diagonal of the basin. A water/sediment mixture can be injected from the top using a peristaltic pump. At the bottom, water was pumped out of the domain such that no standing water would accumulate in the basin. An initial surface was generated inside the flume using a sediment layer with a height of $5cm$ at the inlet and $0cm$ at a distance of about $1.1m$ along the diagonal. As sediment we used crushed glass with a diameter of 50 to 120 microns, which was mixed with water in a tank and then injected into the domain. A description of the setup is shown in Fig.5.9.

The experiment was run in epochs with water/sediment injection followed by a period of drying. After each epoch the surface topography was measured using a 3D scanner, allowing us to obtain coregistered layers of the different sediment facies. With this method it is possible to quantify surface changes and the distribution of sediment. To hasten the drying, an array of fifteen 300W heat lamps were placed $30cm$ above the surface.

Complete drying was necessary to avoid specular reflections produced by the wet sand; these reflections would otherwise disturb the scanning. The scanning technique is based on a stereoscopic measurement where a regular grid is projected onto the surface and the deformation is measured using a camera. From the deformation of the grid lines the topography can be reconstructed.

The scan of the initial surface is shown in Fig.5.11(a) and after the first injection in Fig.5.11(b). The deposition region can be clearly identified.

Figure 5.10 shows the results after 5 epochs of an experiment in which water/sediment was injected for 45 minutes followed by a drying period of 2

CHAPTER 5. INLAND DELTA GENERATION

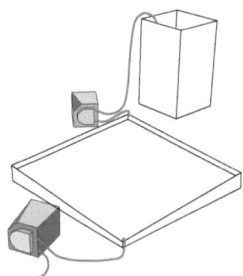 

Fig. 5.9: *Sketch of the experimental setup (left). Water and sediment are fed from a container (background) where the sediment is kept in suspension using an electric mixer. The sediment-water suspension then is injected into the basin using a peristaltic pump (Ismatec ecoline). At the end of the flume, remaining water is pumped out of the domain. Initial condition of the experiment (right).*

Fig. 5.10: *Left: Photo of the experimental setup. The heat lamps are used to dry the sand rapidly. The water/sediment mixture is injected on the left corner. Photo of the 3D scanning device is shown (right).*

5.3. EXPERIMENTAL MODELING

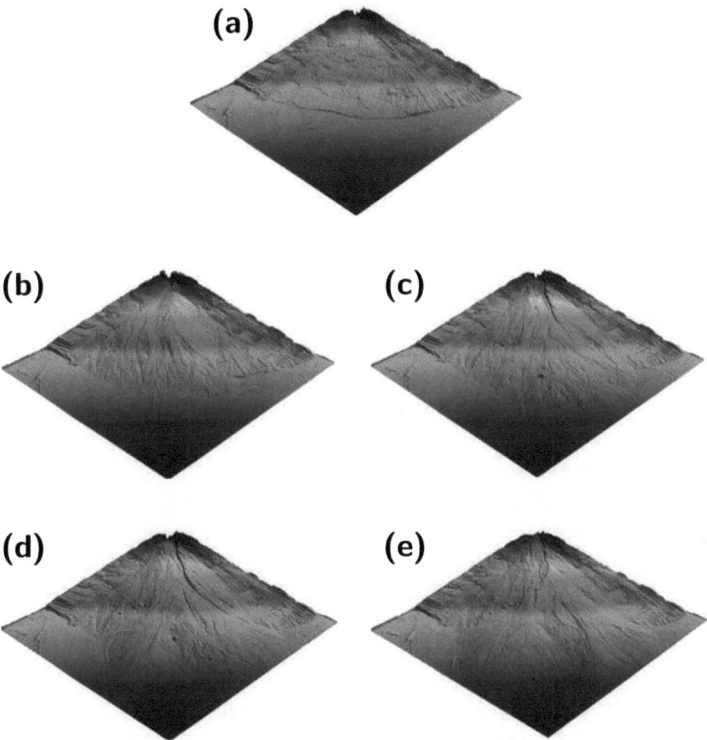

Fig. 5.11: *The figure shows the deposition pattern after the different epochs, where the different lobes can be clearly identified. The initial condition is shown in figure (a).*

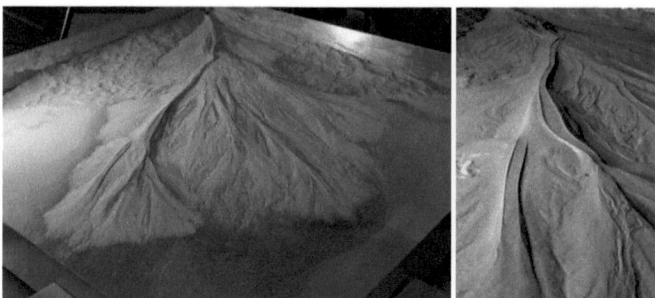

Fig. 5.12: *Deposition pattern of the experiment after several cycles of delta formation (left). When the stream looses its power the channels start to braid. The levees confining the channel are clearly visible (right).*

hours. The volumetric sediment concentration for the run was chosen to be around 0.05 and the inflow is 1000ml/h (pump level 4). The switching of the fan could clearly be identified.

For lower inclines the channels tended to form more distinct levees confining the water flow in a single channel, to the point where they spread out and distribute the sediment on the fan (Fig.5.12).

Although the experimental results cannot be directly rescaled to the natural case, the pattern formation mechanisms and the resulting patterns are similar to those observed in nature. In the small scale experimental set up, the grain size and densities are not scaled with the flow variables and also surface tension effects are predominant.

The total deposited volume of each epoch including pore space can be obtained by simply subtracting the DEM data from the previous one. For the four injections we obtain $V_{0,1} = 573cm^3$, $V_{1,2} = 904cm^3$, $V_{2,3} = 614cm^3$ and $V_{3,4} = 740cm^3$, which gives an estimate of the variance of sediment input during the different runs. Here the change of the porosity due to later compaction by new sediment layers is not taken into account.

In the following we analyze the static and dynamic properties of the deposition pattern and compare the geomorphological features of inland deltas, using the experimental, simulated and real DEM data. For quantifying the geomorphology of the different surface patterns we calculate the radial elevation statistics on circular arcs starting from the inlet. Figure 5.13 shows the mean height of the experimental runnings in a segment of $[d, d + dr]$

5.3. EXPERIMENTAL MODELING

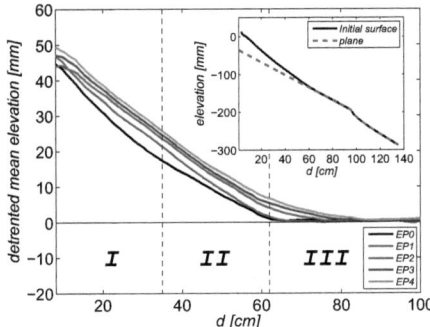

Fig. 5.13: *Statistics of the mean elevation on rings with distance $[d, d + dr]$ from the inlet. The data has been detrended by removing the mean average slope of the base.*

plotted versus the distance d from the inlet. The linear trend of the basin's base has been removed from the data

$$H(d) = \langle H \rangle_{d,d+dr} - \langle B \rangle_{d,d+dr} \quad (5.5)$$

where $H(x, y)$ is the total height and $B(x, y)$ describes the inclined plane corresponding to the base of the basin. In the inset of Fig. 5.13 the total height statistics are presented where the base slope is marked with a dashed red line.

The spatial distribution of the sediment in the domain is shown in Fig. 5.16 where the deposition between successive injections is plotted. The entire deposit of each epoch compared with the initial condition is presented in Fig.5.15. The different lobes can be identified clearly. While in epochs 1 and 2 the central region is filled mainly by pushing forward the central lobe the deposits in epochs 3 and 4 are shifted to the sides, depositing predominantly at the end of their transporting channels which are confined between two distinct levees (Fig. 5.12). Levee formation can also be observed in the Okavango Delta and in the computer simulations. A cross section through one of the anastomosing channel ends of the simulation is shown in Fig.5.14.

The two most common parameters used to characterize a topographic surface are the local slope and the local aspect of the topography. The aspect identifies the steepest downslope direction from each cell to its neighbors. The plots of the local slope and aspects for the different epochs are summarized in Fig.5.17 The local slope is an important characteristic feature in geo-

86   CHAPTER 5. INLAND DELTA GENERATION

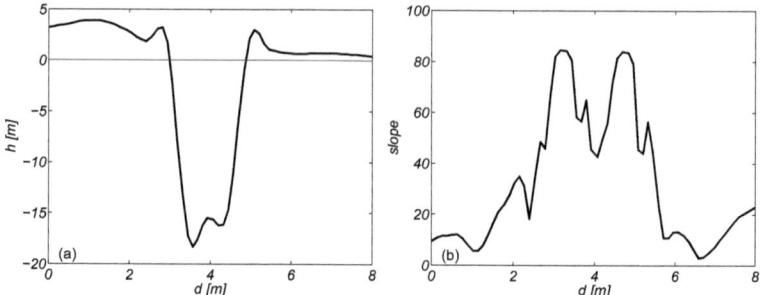

Fig. 5.14: *Cross section through a channel of the simulation including evaporation (a). The levees confining the channel can be clearly identified. The actual water level is marked with a blue line. (b) Slope along the same cross section of the channel.*

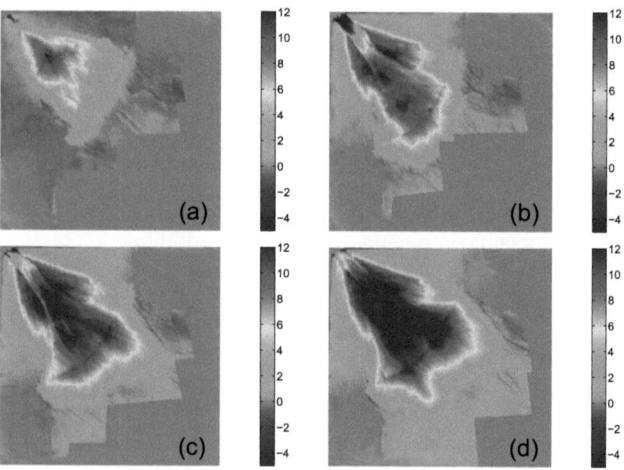

Fig. 5.15: *Deposition pattern of epochs 1-5 with respect to the initial condition. The different deposition areas can be identified.*

5.3. EXPERIMENTAL MODELING

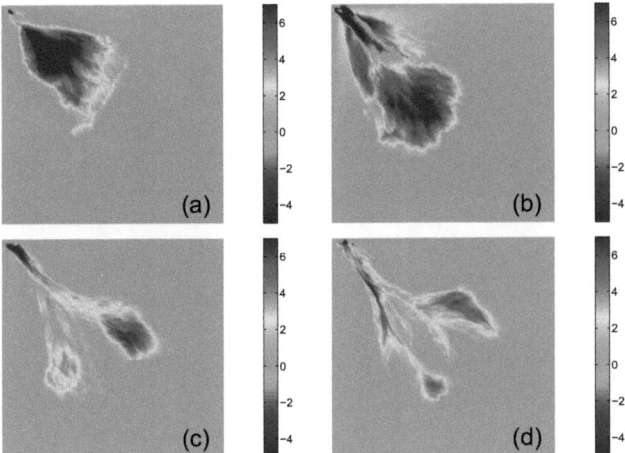

Fig. 5.16: *Deposition pattern of successive epochs.*

morphological analysis, because it gives insight into land forming processes. For example a downstream slope is an indicator of a fluvially dominated surface evolution, while a constant or increasing slope indicates the dominance of hillslope processes. In the fluvial dominated regime the sediment transport capacity Q_s is directly related to the drainage area A and the average slope s, given by the relation

$$Q_S \sim A^m s^n \tag{5.6}$$

where in the delta region the contributing drainage area is always the same and can be set constant. Thus the sediment transport capacity is directly proportional to the slope at some power n around 0.5.

For the different injections we calculate the mean slope $S(d)$ in a ring segment of distance $d+dr$ from the inlet and compare the result with the initial condition. The resulting slopes of the laboratory measurements, the DEM data and the simulation are normalized with the average mean slope of the whole system to obtain dimensionless comparable variables.

$$S(d) = \frac{1}{\langle S \rangle} \langle S \rangle_{d+dr} \tag{5.7}$$

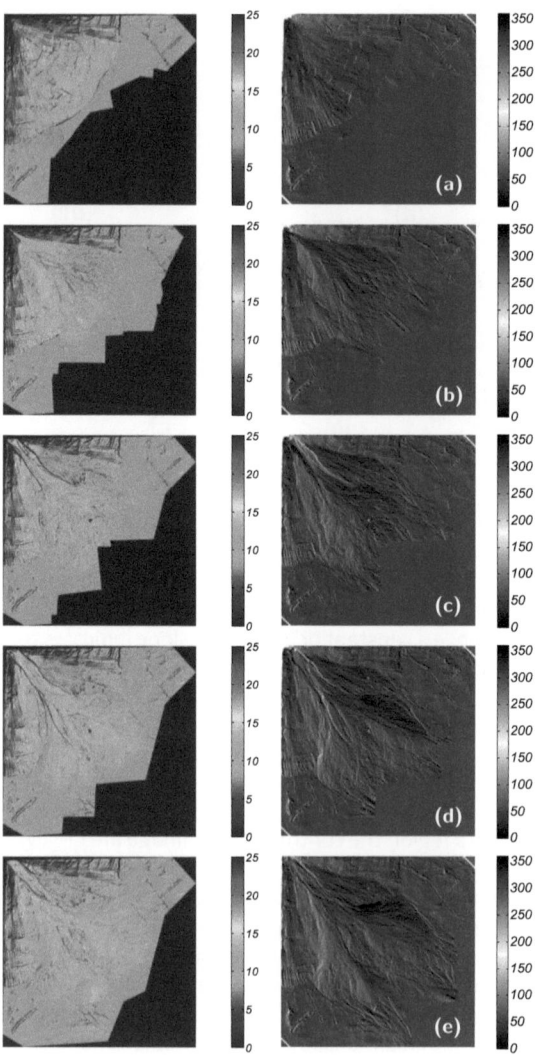

Fig. 5.17: *Slope and local aspect for the different epochs of the experiment; initial condition (a), epoch 1 (b), epoch 2 (c), epoch 3 (d), epoch 4 (e).*

5.3. EXPERIMENTAL MODELING

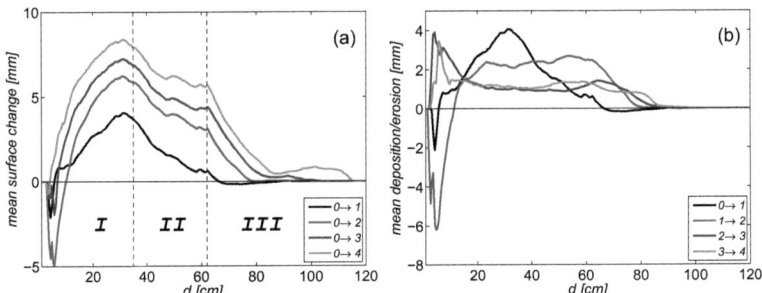

Fig. 5.18: *(a) Plot of the mean change of the elevation vs. distance from the inlet compared to the initial condition. The values have been averaged over a stripe of* $dr = 0.5cm$. *(b) The same plot but comparing successive epochs of the experiment.*

The normalized mean slopes $S(d)$ of the experimental runs are plotted in Figs.5.19(b). One can observe that the slope gradually decreases until the base slope of the flume is reached. With increasing sedimentation the point where this transition happens is shifted forward in the plain. Fig.5.19(a) shows the deviation of the mean slope in a distance d from the initial slope. The downstream change of the topography profile during the experimental run is presented in Figs.5.18(a) and (b) where (a) shows the deposition statistics compared with the initial condition and (b) shows the deposition statistics of successive injections is shown.

The erosive regime close to the inlet and the deposition region at the far part of the flume can be identified. The high changes at the inlet zone are boundary artefacts which are related to the way the water/sediment mixture is injected into the domain. Altogether the whole domain is in a deposition dominated regime, where 3 regions can be distinguished: (I) a region where the sediment is transported in a well-confined channel where the deposition increases with distance from the inlet resulting in a smaller average slope compared to the initial condition (distance 5cm-35cm). (II) The region between 35cm and 60cm marks the range where the flow starts to spread out, distributing the sediment over a larger area of the domain and result in a gradual decrease of deposition while keeping a constant transport capacity Q_s indicated by a constant slope. (III) Towards the end of the delta the flow starts to deposit its final sediment load on the base of the flume in finger like configurations where the small channels are confined by distinct

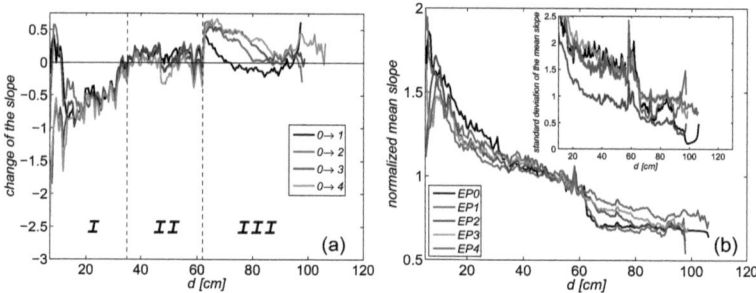

Fig. 5.19: *(a) Plot of the change of the mean local slope vs. distance from the inlet averaged over a stripe of* $dr = 0.5cm$, *(b) main plot: mean local slope as function of the distance from the inlet. The inset shows the standard deviation of the mean local slope depending on the distance d*

levees. This results in an increased average slope compared with the initial condition towards the end of the domain. The three regions are marked in Fig.5.18(a) with (I-III) and Fig.5.19(a).

The variability in slope, expressed by standard deviation statistics (c.f. Fig.5.19(b)), describes the variability of the delta surface topography in the downstream direction. This measure is an indicator for the dynamics of the sedimentation process. Delta switching, local lobe growth, and channel accretion result in a higher standard deviation while uniform gradual deposition over larger areas leads to lower values.

For comparison Fig.5.20 shows the mean slope of the Okavango Delta calculated in radial distances of $r + dr$ from the panhandle inlet.

The calculation of the slope statistics for the Okavango is also shown in Fig.5.20 where in (a) we plotted the change of the mean slope with distance d from the panhandle inlet while the mean standard deviation versus distance from the inlet is plotted in Fig.5.20(b).The DEM data used for the calculation is based on the microtopography model by [71]. During the first $100km$ the Okavango is confined between the fault lines forming the panhandle, which results in a high slope area with decreasing mean slope towards the outlet of the panhandle zone. The delta surface is almost entirely flat with only small variances in the average slope. The large drop in mean slope at the end of the curve marks the end of the Delta which is defined by the Kunyere and Thamalakane Faults.

5.3. EXPERIMENTAL MODELING

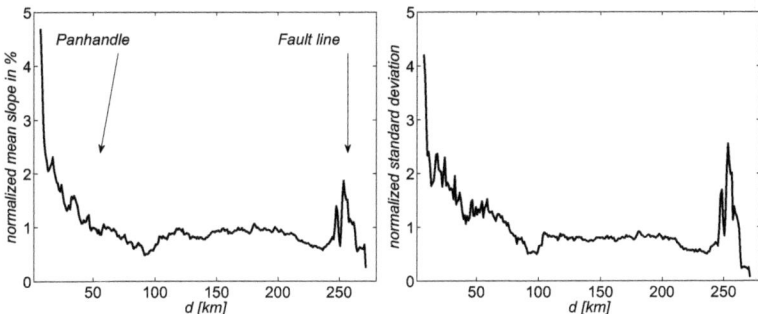

Fig. 5.20: *Plot of the normalized mean local slope of the Okavango delta vs. distance from the inlet averaged over a stripe of* $dr = 1000m$ *(left). Standard deviation depending on the distance d (right).*

For comparison of the dry delta analysis with the simulation we also performed the same statistical analysis on the simulated topography grid. As bases for the analysis we use an interpolated DEM model obtained from the simulated data points. The results are shown in Fig.5.21.

Topographic analysis of the simulation results shows qualitatively comparable behavior to the experiment and the DEM analysis, and it compares better with the experimental results than with the Okavango data. This can be explained by the fact, that both experiment and simulation start with an unconfined almost flat surface, while the Okavango is confined by distinct boundaries defined by the tectonic fault lines.

In contrast to the dry delta case, the analysis of the Mississippi Delta Balize Lobe shows a completely different topographic behavior as shown in Fig.5.22, where the mean slope increases with distance to the inlet. The behavior also is observed in the simulation of a coastal river delta system shown in Fig.5.23 and compares well with the bathymetry measurements. The slope statistics of the delta topography are indicators for sediment mobility in the downstream system. The decreasing slope in the case of inland deltas therefore is consistent with the decreasing sediment transport capacity of the anastomosing channels due to water loss. In the case of coastal deltas, however, the sediment transport capacity of the stream does not change considerably downstream. Only at the mouth, where the river debouches into the ocean, do wave and tidal currents cause high sediment mobility resulting in a steeper slope.

CHAPTER 5. INLAND DELTA GENERATION

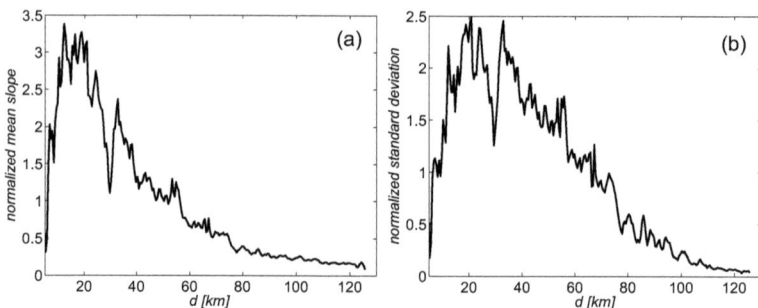

Fig. 5.21: *Plot of the mean local slope vs. distance from the inlet for the inland delta simulation delta(left). The slope has been averaged over a stripe of* $dr = 500m$. *Standard deviation of the mean local slope depending on the distance d (right).*

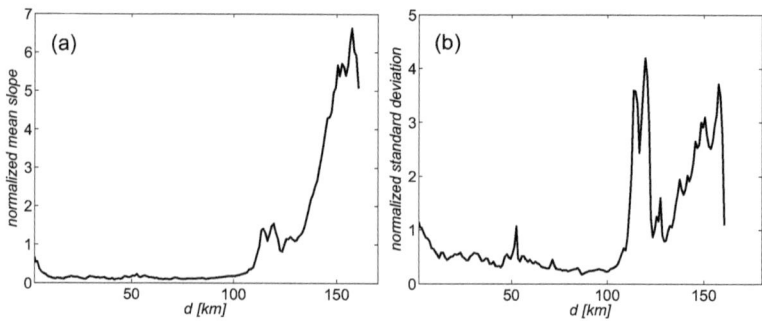

Fig. 5.22: *Right: Plot of the mean local slope of the Balize Lobe in the Mississippi Delta vs. distance from the inlet averaged over a stripe of* $dr = 1000m$. *Right: standard deviation of the mean local slope depending on the distance d.*

5.3. EXPERIMENTAL MODELING

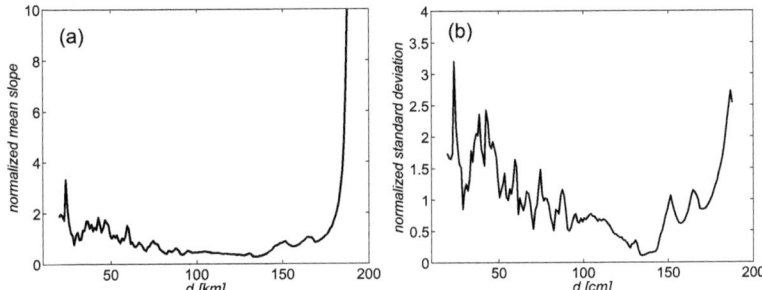

Fig. 5.23: *Left: Plot of the mean local slope vs. distance from the inlet averaged over a stripe of* $dr = 1000$. *Right: For a simulation of a birdfoot delta, standard deviation of the mean local slope depending on the distance d.*

Detailed analysis of the topographic features and comparisons of wet and dry deltas using more sophisticated characterization techniques like the D_8 or D_∞ flow analysis by Tarboton [146] and the calculation of the downflow influence is ongoing research in the SNF project 200021-116050.

Chapter 6

Conclusions

Reduced complexity models have shown to be a promising approach in the geomorphological modeling of fluvial processes. They have proved to be numerically simpler and more suitable for long-term simulation than their detailed, fully physically-based counterparts (e.g., [26; 47; 120; 112; 113; 153]). Recent applications to delta formation (e.g., [142; 137]) have shown that reduced complexity models can capture the essence of the delta formation mechanisms, i.e. the processes of transport, erosion and deposition of sediment by distributary channels. In this work, a new approach for modeling hydrodynamics using a flow resistance and continuity equation has been developed. In contrast to entirely flow-routing based models, the presented model incorporates a well defined water surface, which allows us to capture channel erosion and deposition - as well as the subaqueous growth of the delta, the formation of levees and the dynamics of the delta evolution. The presented model is the first reduced complexity model that captures these features which topography driven models based on Manning-Strickler-like flow routing schemes were not able to reproduce. Especially at the coastline these models fail to capture the subaqueous deposition process because the flow cannot be described accurately by channel flow formulas anymore. The topography is coupled with the flow by a phenomenological sedimentation- erosion law and sediment is advected by the water flow. The presented model incorporates a physically-based flow routing scheme and not hindered by this limitation, thus it shows its strength especially in the formation of a complex subaqueous landscape with channels and levees forming the base of the delta.
The model adequately reproduces several of the static and dynamic phenomena of real deltas, like the fractal surface pattern or the dynamic process of lobe switching, which are an intrinsic properties of the model itself

and not triggered by a parameter. Furthermore, the different delta types – according to the classification scheme of Galloway – can be reproduced with the model by modifying the sedimentation- erosion law.

In Chapter4.5 we go into more detail by applying the model [137] to the simulation of a real river-dominated delta, the Mississippi. The aim is to show that the model is internally consistent and gives physically meaningful results, that the dimensionless parameters may be rescaled to fit observations, and that the model produces long-term simulated dynamics of the delta formation process with a complex temporal correlation structure.

The internal consistency of the model has been vaildated by looking at the space and time-integrated distributions of water and sediment fluxes and sedimentation rates over the delta. The erosion-deposition law in the model is simple; however, it is both sediment supply- and transport capacity-limited, and it produces a plausible distribution of actual sedimentation rates. These rates should be verifiable by measurements of deposition rates in the delta from submarine cores. One of the main advances in this model is that it generates subaerial and subaqueous channels and lateral levee formations, similar to those observed in nature. The shape of the subaqueous part of the simulated delta corresponds accurately to observations of the Mississippi Delta from bathymetric data, which suggests that sediment transport and deposition of sediment are captured well by the model.

The dimensionless parameters of the original model have been rescaled to match the formation of the most recent Balize Lobe of the Mississippi Delta for which we have sufficient data available. Lattice dimensions have been determined by scaling the geometry of the delta. Water and sediment inflow have been rescaled using the observed mean annual water and sediment fluxes, measured at the upstream end of the delta. The modeling time step has been scaled by the estimated age of the delta lobe and mass consistency has been checked by comparing the simulated and observed delta volumes determined from bathymetric data. The result is that the original dimensionless model formulation is cast in dimensional terms, which allows a better appreciation of the process rates and the possibility to examine the scale-dependence of parameters.

Finally, the simulated growth of the subaerial part of a birdfoot delta has been analyzed with the detrended fluctuation analysis technique to assess the nature of long-range temporal correlation in delta growth. The results show that the simulated dynamics of birdfoot delta growth have a characteristic timescale with a transition from a highly correlated regime for small timescales, to a less correlated one at larger timescales. We hypothesize that the characteristic timescale separates periods of consistent delta growth by gradual sediment deposition at the mouths of distributary channels from periods, at which random large scale channel avulsions lead to rapid change and the formation of new channels and subaqueous dominated deposition. The simulated temporal dynamics cannot be directly verified by observations of the Mississippi Delta surface, but intuitively they do appear correctly. Apart from collecting data to verify the modeled process rates and time reconstructions in the Mississippi, future work should be aimed at the examination of the resolution dependence of the model parameters, their generality for river-dominated deltas, and the sensitivity of the results to the model parameter set. The erosion-deposition law, which in the current model is flow rate dependent and contains one parameter which represents both the erodibility of the surface as well as the trapping efficiency of fine material, may also be expanded by considering erosion and deposition separately. Information on grain size distributions and incipient motion thresholds can be added as well.

In Chapter 5 the model is extended to simulate the dynamics of inland delta generation including evaporation and seepage in the hydrodynamics. It has been observed in this application that the model is limited in reproducing channel flow correctly and that Manning-Strickler based models perform better in capturing the channel dynamics.

For comparison of the model results with real landforms we have chosen the Okavango Delta, which is one of the largest inland alluvial fans in the world. The simulation has been accompanied by a flume experiment to study the switching behavior of the fan formation and the layering of the different sediment facies. The topographic structure of the fan has been analyzed and conpared with the simulation and the real Okavango micro-topography. In ongoing studies the characterization and comparison of inland deltas with coastal deltas will be investigated to characterize and quantify the difference between dry and wet deltas.

Further development of the model will include vegetation and ground water flow and salt transport, which play an important role in the pattern processes of the Okavango.

Acknowledgment

I wish to express my sincere gratitude to Prof. H.J. Herrmann who gave me the opportunity to work at ETH Zurich, offered me excellent working conditions, and continuous encouragement throughout this work.
I also thank those people who contributed in this work, especially Prof. J.S. Andrade who assisted the work throughout the project with scientific advice and comments. Furthermore, I would like to thank my co-advisors Prof. W. Kinzelbach and Prof. H. Seyfried for carefully correcting the thesis, and Dr. P. Molnar for advices and help in the SNF project 200021-116050 and Dr. D. Akça for the scanning of the experiment.
Finally I would like to thank my parents and Evelyn Strombom for proofreading the text.
I also am grateful to the members of IFB, ETH Zurich who indirectly supported this work by an excellent working atmosphere and supported me with help during the experimental setup.
This work was funded by the Swiss National Fonds No 200021-116050.

Hansjörg Seybold
Zurich, July 7, 2010

References

[1] Rockware, Earth Science and GIS software *http://www.rockware.com*

[2] Boss International *http://www.bossintl.com/*

[3] The CCHE2D Model. University of Mississippi,Carrier Hall, Room 102 University, MS 38677

[4] The CCHE3D Model. University of Mississippi,Carrier Hall, Room 102 University, MS 38677

[5] F. Ahnert. Brief description of a comprehensive three-dimensional process-response model of landform development. *Zeitschrift für Geomorphologie*, 25:29–49, 1976. Suppl.

[6] G.P. Allen. Costal geomorphology of eastern nigerian beach ridge barrier island and vegetated tidal flats. *Geol. en Mijnb*, 44(1):1–21, 1965.

[7] G.P. Allen, D. Laurier, and J. P. Thouvenin. Modern Mahakam Delta, Indonesia - sand distribution and geometry in mixed tide and fluvial delta. *A.A.P.G. Bull.*, 65(5):889–889, 1981.

[8] P.E. Ashmore. Laboratory modelling of gravel braided stream morphology. *Earth Surf. Proc. Land.*, 7:201–225, 1982.

[9] P.E. Ashmore. *Process and Form of Graval Braided Streams: Laboratory Modelling and Field Observations*. PhD thesis, University of Alberta, Alberta, 1985.

[10] A.B. Atekwana, J.P. Hogan, A.B. Kampunzu, and M.P. Modise. Early structural evolution of the nascent Okavango Rift Zone, N.W. Botswana. In *East African Rift System Conference*, Addis Ababa, Ethiopia, June 2004.

[11] Z Babinski. The relationship between suspended and bed load transport in river channels. In D. E. Walling and A. J. Horowitz, editors, *Sediment budgets 1*, volume 291, pages 182–188, 2005.

[12] C.C. Bates. Rational theory of delta formation. *A.A.P.G. Bull.*, 37(9):2119–2162, 1953.

[13] P.D. Bates and A.P.J. de Roo. A simple raster-based model for floodplain inudation. *J. Hydrol.*, 236:54–77, 2000.

[14] P. Bauer. *Flooding and salt transport in the Okavango Delta, Botswana*. PhD thesis, ETH Zürich, 2004.

[15] P. Bauer, T. Gumbricht, and W. Kinzelbach. A regional coupled surface water/groundwater model of the Okavango Delta, Botswana. *Water Res. Res.*, 42, 2006.

[16] P. Bauer, G. Thabeng, F. Stauffer, and W. Kinzelbach. Estimation of the evapotranspiration-rate from diurnal groundwater level fluctuations in the okavango delta, botswana. *Journal of Hydrology*, 288(3-4):344–355, 2004.

[17] C. Beaumont, H. Kooi, and C. Willett. Coupled tectonic-surface process models with applications to rifted margins and collisional orogens. In M.A. Summerfield, editor, *Geomorphology and Global Tectonics*, pages 29–55. Wiley, 2000.

[18] H.A Bernard. A resume of delta types. *Am. Ass. Petrol. Geol. Bull.*, 49:334–335, 1965.

[19] H.A Bernard and Le Blanc R.J. *Resume of Quaternary geology of the northwestern Gulf of Mexico provinece*, pages 137–185. Princton Univ. Press, 1965.

[20] K.J. Beven and Kirkby M.J. A physically based, variable contributing area model for basin hydrology. *Hydrol. Sci. B.*, 24(1):43–69, 1979.

[21] J. P. Bhattacharya and R. G. Walker. Deltas. In R. G. Walker and N.P. James, editors, *Facies Models, Response to Sea-Level Change*, pages 157–177. Geological Association of Canada, St. Johns, 1992.

[22] R. Bleck. Finite-difference equations in generalized vertical coordinates. part i:total energy conservation. *Contrib. Atmos. Phys.*, 51:360–372, 1978.

[23] R. Bleck. Simulation of coastal upwelling frontogenesis with an isopycnic coordinate model. *J. Geophys. Res*, 83:6163–6172, 1978.

[24] R. Bleck. Finite-difference equations in generalized vertical coordinates. part ii: Potential vorticity conservation. *Contrib. Atmos. Phys.*, 52:95–105, 1979.

[25] A. F. Blumberg and G.L.. Mellor. A description of a three-dimensional coastal ocean circulation model. In N.S. Heaps, editor, *Three-Dimensional Coastal Ocean Models*, pages 1–16. American Geophysical Union, Washington, D.C., 1987.

[26] J. Brasington and K. Richards. Reduced-complexity, physically-based geomorphological modelling for catchment and river management. In *Geomorphology, special issue*, volume 90, pages 171–177. Elsevier, 2007.

[27] R.W.G Carter and C.D. Woodroffe. *Coastal Evolution*. Press Syndicate of the University of Cambridge, 1994.

[28] O. Catuneanu. *Principles of Sequence Stratigraphy*. Elsivir, 2006.

[29] Vadim V. Cheianov, Vladimir I. Fal'ko, Boris L. Altshuler, and Igor L. Aleiner. Random resistor network model of minimal conductivity in graphene. *Physical Review Letters*, 99(17):176801, 2007.

[30] C. Chen, R.C. Beardsley, and G. Cowles. An unstructured grid, finite-volume coastal ocean model (FVCOM) system. *Oceanography*, 19(1):78–89, 2006.

[31] Zhi Chen, Plamen Ch. Ivanov, Kun Hu, and H. Eugene Stanley. Effect of nonstationarities on detrended fluctuation analysis. *Phys. Rev. E*, 65(4):041107, Apr 2002.

[32] J.M. Coleman. *Deltas: Processes of Deposition and Models for Exploaration*. Continuing Education Publication Company, Inc., Campain, IL, 1975.

[33] J.M. Coleman. Dynamic changes and processes in the Mississippi River Delta. *Geol. Soc. Am. Bull.*, 100:999–1015, 1988.

[34] J.M. Coleman and S.M. Gagliano. Cyclic sedimentation in the Mississippi River Delta Plane. *Gulf Coast Assn. Geol. Soc. Trans.*, 14:67–80, 1964.

[35] J.M. Coleman and S.M. Gagliano. Cyclic sedimentation in the mississippi river delta plane. *Gulf Coast Assn. Geol. Soc. Trans.*, 14:67–80, 1964.

[36] J.M. Coleman and S.M. Gagliano. Sedimentary structures: Mississippi river deltaic plain. In G. V. Middleton, editor, *Primary sedimentary structures and their hydrodynamic interpretation*, volume 21, pages 133–148. SEMP Special Publication, 1965.

[37] J.M. Coleman and D.B. Prior. *Deltaic Sand Bodies*. Number 15 in Continuing Education, Course Note Series. A.A.P.G., 1980.

[38] J.M. Coleman and L.D. Wright. Variability of modern river deltas. *Trans. Gulf Coast Assoc. Geol. Soc.*, 23:33–36, 1973.

[39] J.M. Coleman and L.D. Wright. Modern river deltas: variability of processes and sand bodies. In M.L. Broussard, editor, *Deltas*, pages 99–149. Houston Geological Society, 1975.

[40] D. R. Corbett, B. McKee, and M. Allison. Nature of decadal-scale sediment accumulation on the western shelf of the Mississippi River delta. *Cont. Shelf Res.*, 26(17-18):2125–2140, 2006. Special Issue in Honor of Richard W. Sternberg's Contributions to Marine Sedimentology.

[41] A. Correggiari, A. Cattaneo, and F. Trincardi. The modern po delta system: Lobe switching and asymmetric prodelta growth. *L. Marin. Geol.*, 222-223:49–74, 2005.

[42] T. J. Coulthard. Landscape evolution models: a software review. *Hydrological Processes*, 15(1):165–173, 2001.

[43] T. J. Coulthard. Effects of vegetation on braided stream pattern and dynamics. *Water Resour. Res.*, 41(4), 2005. 04003, doi:10.1029/2004WR003201.

REFERENCES

[44] T. J. Coulthard, M. J. Kirkby, and M. G. Macklin. Non-linearity and spatial resolution in a cellular automaton model of a small upland basin. *Hydrology and Earth System Sciences*, 2(2/3):257–264, 1998.

[45] T. J. Coulthard and M. J. van de Viel. A cellular model of river meandering. *Earth. Surf. Proc. Land.*, 31(1):123–132, 2006.

[46] T.J. Coulthard, D.M. Hicks, and M.J. Van De Wiel. Cellular modelling of river catchments and reaches: Advantages, limitations and prospects. *Geomorphology*, 90(3-4):192 – 207, 2007. Reduced-Complexity Geomorphological Modelling for River and Catchment Management.

[47] A. Crave and P. Davy. A stochastic "precipiton" model for simulating erosion/sedimentation dynamics. *Computers & Geosciences*, 27(7):815 – 827, 2001.

[48] A. Czirók and E. Somfai. Experimental evidence for self-affine roughening in a micromodel of geomorphological evolution. *Phys. Rev. Lett.*, 71(13):2154–2157, 1993.

[49] P. Davy and A. Crave. Upscaling local-scale transport processes in large-scale relief dynamics. *Physics and Chemistry of the Earth, Part A: Solid Earth and Geodesy*, 25(6-7):533 – 541, 2000.

[50] A.L. Densmore, R.S. Anderson, B.G. McAdoo, and M.A. Ellis. Hillslope evolution by bedrock landslides. *Science*, 275:369–372, 1997.

[51] W.E. Dietrich, G. Day, and Parker. G. The fly river, papua new guinea: Inferences about river dynamics, floodplain sedimentation and fate of sediment. In A.J. Miller and A. Gupta, editors, *Varieties in Fluvial Form*, pages 345–376. John Wiley, 1999.

[52] T. Dincer, L.G. Hutton, and B.B.J. Khupe. Study, using stable isotopes of flow distribution, surface-ground water relations and evapotranspiration in the okavango delta, botswana. *Int. Atomic Energy Agency Ser. STI Pub.*, 493:3–26, 1981.

[53] D.L. Divins and D. Metzger. Coastal relief model, Central Gulf of Mexico Grids, 2006.

[54] A.B. Doeschel-Wilson and P.E. Ashmore. Assessing a numerical cellular braided-stream model with a physical model. *Earth. Surf. Proc. Land.*, 30:519–540, 2005.

[55] A. E. Draut, G. C. Kineke, D. W. Velasco, M. A. Allison, and R. J. Prime. Influence of the Atchafalaya River on recent evolution of the chenier-plain inner continental shelf, northern Gulf of Mexico. *Cont. Shelf Res.*, 25(1):91–112, 2005.

[56] G. Duan. *Simulation of Alluvial Channel Migration Processes with a Two-dimensional Numerical Model*. PhD thesis, University of Mississippi, Oxford, Mississippi, 1998.

[57] D. A. Edmonds and R. L. Slingerland. Mechanics of river mouth bar formation: Implications for the morphodynamics of delta distributary networks. *J. of Geophys. Res. Earth-Surface*, 112(F03099):14, 2007.

[58] D. A. Edmonds and R. L. Slingerland. Stability of delta distributary networks and their bifurcations. *Water Res. Res.*, 44(9):13, 2008. W09426.

[59] C. Ehrat. *Modelling of Sediment Transport with Application to the Okawango Delta*. PhD thesis, ETH Zürich, 2006.

[60] K Ellery, W.N. Ellery, K.H. Rogers, and B.H. Walker. Formation, colonization and fate of floating sudds in the mainachira river system of the okavango delta, botswana. *Aquatic Botany*, 38:315–329, 1990.

[61] J. Feder. *Fractals*. Plenum Press, New York, 4 edition, 1989.

[62] H.N. Fisk. Geological investigation of the alluvial deposits and their effects on mississippi river activity. Technical report, U.S. Army Corps of Engineering, Mississippi River Commission, 1944.

[63] H.N. Fisk. Fine-grained alluvial deposits and their effects on mississippi river activities. Technical report, U.S. Army Corps of Engineering, Mississippi River Commission, 1947.

[64] H.N. Fisk. Geological investigation of the atchafalaya basin and the problem of mississippi river diversion. Technical report, U.S. Army Corps of Engineering, Mississippi River Commission, 1952.

[65] H.N. Fisk. Bar-finger sands of the mississippi deltacretaceous sedimentation in the upper mississippi embayment. In *Geometry fo sandstone bodies*, volume 44, pages 29–52. Am. Assoc. Petr. Geol., 1961.

REFERENCES

[66] G. R. Foster and L. D. Meyer. A closed-form erosion equation for upland areas. *Sedimentation: Symposium to Honor Prof. H. A. Einstein*, pages 12.1–12.19, 1972.

[67] W.E. Galloway. Process framework for describing the morphologic and stratigraphic evolution of deltaic depositional systems. In M.L. Broussard, editor, *Deltas*, pages 87–98. Houston Geological Society, 1975.

[68] R.L. Gawthrop, S. Hardy, and B. Ritchie. Numerical modelling of depositioan sequences in half-graben rift basins. *Sedimentology*, 12:1083–1086, 2003.

[69] A. Giacometti, A. Maritan, and J.R. Banavar. Continuum model for river networks. *Phys. Rev. Lett.*, pages 577–580, 1995.

[70] L. Giosan and J.P. Bhattacharya, editors. *River Deltas- Concepts, Models and Examples*. SEMP, 2005.

[71] T. Gumbricht, J. McCarthy, and T. S. McCarthy. Channels, wetlands and islands in the okavango delta, botswana, and their relation to hydrological and sedimentological processes. *Earth Surface Processes and Landforms*, 29(1):15–29, 2004.

[72] R. Hallberg. Some aspects of the circulation in ocean basins with isopycnals intersecting the sloping boundaries. Technical report, University of Washington, 1995.

[73] A. B. Harris and R. Fisch. Critical behavior of random resistor networks. *Phys. Rev. Lett.*, 38(15):796–799, Apr 1977.

[74] C.K. Harris, P.A. Traykovski, and W.R. Geyer. Flood dispersal and deposition by near-bed gravitational sediment flows and oceanographic transport: A numerical modeling study of the Eel river shelf, northern california. *J. Geophys. Res. Oceans*, 110(C09025):1–16, 2005.

[75] A. D. Howard. A detachment-limited model of drainage basin evolution. *Water Resour. Res.*, 30(7):2261–2285, 1994.

[76] D.C. Hoyal and B.A. Sheets. Morphodynamic evolution of cohesive experimental deltas. *J. Geophys. Res. Earth-Surface*, 114(F02009):78–89, 2009.

[77] Kun Hu, Plamen Ch. Ivanov, Zhi Chen, Pedro Carpena, and H. Eugene Stanley. Effect of trends on detrended fluctuation analysis. *Phys. Rev. E*, 64(1):011114, Jun 2001.

[78] E.W. Hurst, R. P. Black, and Y. M. Simaika. *Long-Term Storage: An Experimental Study*. "Constable & Co, London, UK", 1965.

[79] D.G. Hutchins, S.M. Hutton, and C.R. Jones. The geology of the okavango delta. In *Symposium on the Okavango Delta*, pages 13–20, Gaborne, 1976. Botswana Society.

[80] T. A. Jaggar. Experiments illustrating erosion and sedimentation. *Bull. Mus. Compar. Zool.*, 49:285–305, 1908.

[81] D.J. Jerolmack and C. Paola. Complexity in cellular model river avulsion. *Geomorphology*, 91(3-4):259–270, 2007. 38th Binghamton Geomorphology Symposium: Complexity in Geomorphology.

[82] Coleman J.M. and Wright L.D. Analysis of major river systems and their deltas: procedures and rationale, with two examples. Coastal studies series, Louisiana State Univ., 1971.

[83] M. P. Keown, A. A. Jr. Dardeau, and E. M. Caussey. Historic trends in the sediment flow regime of the mississippi river. *Water Resour. Res.*, 22(11):1555–1564, 1986.

[84] L. Kgotlhang. *Application of airborne geophysics in large scale hydrological mapping; Okavango Delta, Botswana*. PhD thesis, ETH Zurich, Zurich, 2008.

[85] W. Kim, C. Paola, V.R. Voller, and J.B. Swenson. Experimental measurement of the relative importance of controls on shoreline migration. *J. Sediment. Res.*, 76(2):270–283, 2006.

[86] M.J. Kirkby. Hillslope process-response models based on the continuity equation, 1971. Special Publication 3.

[87] C. R. Kolb and J.R. van Lopik. Geology of the mississippi deltaic plain, southeastern louisiana. Technical report, U.S. Army Corps of Engineering, Waterways Experiment Station, 1958.

[88] C.R. Kolb and J.R. van Lopik. Depositional environments of the mississippi river deltaic plain-southeastern louisianathe fly river, papua new guinea: Inferences about river dynamics, floodplain sedimentation and fate of sediment. In *Deltas and their Geological Framework*, pages 17–61. Huston Geol. Soc., 1966.

[89] H. Kooi and C. Beaumont. Large-scale geomorphology: Classical concepts reconciled and integrated with contemporary ideas via a surface processes model. *J. Geophys. Res.*, 101:3361–3386, 1996.

[90] E.R. Kraal, M. van Dijk, G. Postma, and M.G. Kleinhans. Martian stepped-delta formation by rapid water release. *Nature*, 451:973–976, 2008.

[91] Ok Tedi Mining LTD. Sixth supplemental agreement environmental study. Technical Report I-III, Ok Tedi Mining LTD, November 1988.

[92] Ok Tedi Mining LTD. Suplementary investigations report. Technical Report I-III, Ok Tedi Mining LTD, November 1989.

[93] D.R. Lynch and F.E. Werner. 3-d hydrodynamics on finite elements, part i: linearized harmonic models. *Int. J. of Num. Methods in Fluids*, 7:871–909, 1987.

[94] D.R. Lynch and F.E. Werner. 3-d hydrodynamics on finite elements, part ii: Nonlinear time stepping models. *Int. J. of Num. Methods in Fluids*, 12:507–533, 1991.

[95] A. Maritan, F. Colaiori, A. Flammini, M. Cieplak, and J.R. Banavar. Universality classes of optimal channel networks. *Science*, 272(984), 1996.

[96] J. Marshall, C. C. Hill, L. Perelman, and A. Adcroft. Hydrostatic, quasi-hydrostatic, and nonhydrostatic ocean modeling. *J. Geophys. Res.*, 102:5733–5752, 1997.

[97] Z. R. P. Mateo and F.P. Siringan. Tectonic control of high-frequency holocene delta switching and fluvial migration in lingayen gulf bayhead, northwestern phillipines. *J. Coast. Res.*, 23:182–194, 2007.

[98] P.A. Matrinez and J.W. Harbaugh. *Simulating nearshore environments*, page 265ff. Pergamon Press, 1993.

[99] T. S. McCarthy, W. N. Ellery, and L. G. Stanistreet. Lakes of the north eastern region of the okavango swamps, botswana. *Zeitschrift Fur Geomorphologie*, 37(3):273–294, 1993.

[100] T. S. McCarthy, K. H. Rogers, I. G. Stanistreet, W. N. Ellery, B. Cairncross, K. Ellery, and T.S.A. Gorbicki. Features of channel margins in the okavango delta. *Palaeoecol. Africa*, 19:3–14, 1988.

[101] T. S. McCarthy, I. G. Stanistreet, and B. Cairncross. The sedimentary dynamics of active fluvial channels on the okavango fan, botswana. *Sedimentology*, 38(3):471–487, 1991.

[102] T.S. McCarthy. Groundwater in the wetlands of the Okavango Delta, Botswana, and its contribution to the structure and function of the ecosystem. *J. Hydrol.*, 320:264–282, 2006.

[103] T.S. McCarthy, W.N. Ellery, and A. Bloem. Some observations on the geomorphological impact of the hippopotamus (hippopotamus amphibius l.) in the Okavango Delta, Botswana. *Afr. J. Ecol.*, 36:44–56, 1998.

[104] X.D. Meijer. Modeling the drainage evolution of a river-shelf system forced by quaternary glacio-eustasy. *Basin Res.*, 14:361–378, 2002.

[105] E. Meyer-Peter and R. Muller. Formulas for bed-load transport. In *Proc. of the 2nd Meeting Int. Association Hydr. Res.*, pages 39–64, Stockholm, Sweden, 1948.

[106] A.D. Miall. Whither stratigraphy. *Sedimentary Geology*, 100:5–20, 1995.

[107] C.. Milzow. *Hydrological and sedimentological modelling of the Okavango Delta Wetlands, Botswana*. PhD thesis, ETH Zurich, Zurich, 2008.

[108] C. Milzow, L. Kgotlhang, W. Kinzelbach, P. Meier, and P. Bauer-Gottwein. The role of remote sensing in hydrological modelling of the okavango delta, botswana. *J. Environ. Manage.*, x:xxx–xxx, 2008.

[109] M.P. Modise, A.B. Atekwana, A.B. Kampuzunu, and T.H. Ngwisanyi. Rift kinematics during the incipient stages of continental extension: evidence from the nascent okavango rift basin, northwest botswana. *Geol.*, 28:939–942, 2000.

REFERENCES

[110] J.P. Morgan. *The Mississippi River Delta, Legal-Geomorphologic evaluation of historic shoreline changes.* Geoscience and Man. Lousiana State University, Baton Rouge, 1977.

[111] J. Mossa. Sediment dynamics in the lowermost Mississippi River. *Eng. Geol.*, 45(1-4):457–479, 1996.

[112] A.B. Murray and C. Paola. A cellular model of braided rivers. *Nature*, 371:54–57, 1994.

[113] A.B. Murray and C. Paola. Properties of a cellular braided stream model. *Earth. Surf. Proc. Land.*, 22:1001–1025, 1997.

[114] The Naval Research Laboratory Ocean Model, NLOM *http://www7320.nrlssc.navy.mil/*

[115] T. Off. Rythmic linear sand bodies caused by tidal currents. *A.A.P.G. Bull.*, 47:324–341, 1063.

[116] E. Oomkens. Lithofacies relations in the late Quaternary Niger Delta complex. *Sedimentology*, 21:145–222, 1974.

[117] G. J. Orton and H. G. Reading. Variability of deltaic processes in terms of sediment supply, with part icular emphasis on grain-size. *Sedimentology*, 40(3):475–512, 1993.

[118] I. Overeem, P.M. Sylvitski, and W.H. hutton. Three-dimensional numerical modeling of deltas. In L. Giosan and J.P. Bhattacharya, editors, *River Deltas- Concepts, Models and Examples*, pages 13–30. SEMP, 2005.

[119] N. Panin, N. Herz, and J.F. Noakes. Radiocarbon dating of danube delta deposits. *Quarternary Res.*, 19:249–255, 1983.

[120] C.J. Paola, C. Mullin, D.C. Ellis, J.B. Mohrig, G. Swenson, T. Parker, P.L. Hickson, L. Heller, J. Pratson, B. Syvitski, B.A. Sheets, and N. Strong. Exprimental stratigaphy. *GSA Today*, 11(7):4–9, 2001.

[121] G. Parker. Sediment transport morphodynamics, with applications to fluvial and subaqueous fans and fan-deltas. edited, copyrighted e-book, 2005.

[122] G. Parker, C. Paola, K. X. Whipple, and D. Mohrig. Alluvial fans formed by channelized fluvial and sheet flow. i: Theor y. *Journal of Hydraulic Engineering-Asce*, 124(10):985–995, 1998. Times Cited: 28.

[123] G. Parker, M. Toro-Escobar, M. Ramey, and S. Beck. The effect of floodwater extraction on mountain stream morphology. *J. Hydraul. Eng.*, 129(11):885–895, 2003.

[124] P. Passalacqua, F. Porté-Agel, E. Foufoula/Georgiou, and C. Paola. Fluvial fan deltas: Linking channel processes with large-scale morphodynamics. *Water Res. Res.*, 42, 2006.

[125] C. K. Peng, S. V. Buldyrev, S. Havlin, M. Simons, H . E. Stanley, and A. L. Goldberger. Mosaic organization of dna nucleotides. *Physical Review E*, 49(2):1685–1689, 1994.

[126] S. Penland, J. R. Suter, and R.A. McBride. Delta plain development and sea level history in the terrebonne costal region, Lousiana. In *Costal Sediments*, pages 1689–1705, 1987.

[127] W.A. Pryor. Cretaceous sedimentation in the upper mississippi embayment. *Am. Assoc. Petr. Geol. bull.*, 44:1473–1504, 1960.

[128] B.D. Ritchie, S. Hardy, and R.L. Gawthorpe. Three-dimensional numerical modeling of coarse-grained lastic deposition in sedimentary basins. *J. Geophys. Res.*, 13:605–627, 1999.

[129] H. H. Roberts. Dynamic changes of the Holocene Mississippi River delta plain: The delta cycle. *J. Coastal Res.*, 13(3):605–627, 1997.

[130] I. Rodriguez-Iturbe and A. Rinaldo. *Fractal River Basins: Chance and Self-Organization*. Cambridge University Press, New Youk, 1997.

[131] B. Sapoval, A. Baldassarri, and A. Gabrielli. Self-stabilized fractality of seacoasts through damped erosion. *Phys. Rev. Lett.*, 2004.

[132] R. J. Saucier. Recent geomorphic history of the Ponchartrain Basin. Technical report, Lousiana State University, 1963.

[133] R. J. Saucier. Geomorphology and quaternary geologic history of the lower mississippi valley e. Technical report, Mississippi River Commision, 1994.

[134] J.M. Schoorl, M.P.W Sonneveld, and A. Veldkamp. Three dimensional landscape process modelling: wave versus flood influence, the effect of dem resolution. *Earth. Surf. Proc. Land.*, 25:1025–134, 2000.

# REFERENCES

[135] P.C. Scruton. Delta building and deltaic sequence. In *Recent sediments, northwest Gulf of Mexico*, pages 82–102. Am. Assoc. Petr. Geol., 1960.

[136] The Rutgers Spectral Element Ocean Model, SEOM Rutgers University *http://marine.rutgers.edu/po/index.php?*

[137] H.J. Seybold, J. S. Andrade, and H. J. Herrmann. Modeling rinver delta formation. *Proc. Natl. Acad. Sci. USA*, 104(43):16804–16809, 2007.

[138] B.A. Sheets, T.A. Hickson, and C. Paola. Assembling the stratigraphic record: Depositional patterns and time-scales in an experimental alluvial basin. *Basin Res.*, 14(3):287–301, 2002.

[139] P.A. Smith. An outline of the vegetation of the okavango drainage system. In *Symposium on the Okavango Delta*, pages 93–112, Gaborne, 1976. Botswana Society.

[140] L. Spencer. *The Fly Estuarine Delta, Gulf of Papua*. PhD thesis, University of Sydney, 1978.

[141] I. G. Stanistreet, B. Cairncross, and T. S. McCarthy. Low sinuosity and meandering bedload rivers of the okavango fan - channel confinement by vegetated levees without fine sediment. *Sedimentary Geology*, 85(1-4):135–156, 1993.

[142] T. Sun, C. Paola, G. Parker, and P. Meakin. Fluvial fan deltas: Linking channel processes with large-scale morphodynamics. *Water Res. Res.*, 38(8), 2002.

[143] C. Swenson, C. Paola, L. Pratson, V. R. Voller, and A. B. Murray. Fluvial and marine controls on combined subaerial and subaqueous delta progradation: Morphodynamic modeling of compound-clinoform development. *J. Geophys. Res.*, 110(F02013):1–16, 2005.

[144] J. P. M. Syvitski, C. J. Vorosmarty, A. J. Kettner, and P. Green. Impact of humans on the flux of terrestrial sediment to the global coastal ocean. *Science*, 308(5720):376–380, 2005.

[145] Surface water modeling system.

[146] D.G. Tarboton. A new method for the determination of flow directions and upslope areas in grid digital elevation models. *Water Res. Res.*, 33(2):309–319, 1997.

[147] D.J. Taylor. A preliminary investigation of the marine geology of the fly river estuary, papua. Technical report, University of Sydney, Department of Geology and Geophysics, 1973.

[148] R. Thomas and A.P. Nicholas. Simulation of braided river flow using a new cellular routing scheme. *Geomporphology*, 43(3-4):179–195, 2002.

[149] R. Thomas, A.P. Nicholas, and T.A. Quine. Cellular modelling as a tool for interpreting historic braided river evolution. *Geomporphology*, 90:171–177, 2007.

[150] G.E. Tucker, S.T. Lancaster, N.M. Gasparini, and R.L. Bras. The channel-hillslope integrated landscape development (child) model. In R.S. Harmon and W.D. Doe, editors, *Landscape Erosion and Sedimentation Mideling*, pages 349–388. Kluwer Academic, 2001.

[151] D.L. Turcotte. *Fractals and chaos in geology and geophysics*. Camebridge University Press, 1997.

[152] Louisiana coastal wetlands restoration plan november 1993. Technical report, USGS, November 1993.

[153] M. J. Van De Wiel, T. J. Coulthard, M. G. Macklin, and J. Lewin. Embedding reach-scale fluvial dynamics within the caesar cellular automaton landscape evolution model. *Geomorphology*, 90(3-4):283–301, 2007.

[154] J. van der Knijff and A. de Roo. Lisflood, distributed water balance and flood simulation model. Technical report, European Union, 2008.

[155] L.C. van Rijn. Handbook of sediment transport by currents and waves. Technical report H461, Delft Hydrolics, 1989.

[156] L.C. van Rijn. *Principles of sediment transport rivers, estuaries and coastal seas*. aquapublications, 1993.

[157] G. Willgoose, R.L. Bras, and I. Rodriguez-Iturbe. A coupled channel network growth and hillslope evolution model, 1, theory. *Water Resour. Res.*, 27(7):1671–1684, 1991.

[158] B.H. Wilson. Some natural and man-made changes in the channels of the okavango delta. *Botswana Notes and Records*, 5:132–153, 1973.

[159] B.H. Wilson and T. Dincer. An introduction to the hydrology and hydrography of the okavango delta. In *Symposium on the Okavango Delta*, pages 33–48, Gaborne, 1976. Botswana Society.

[160] K.X. Wipple and G.E. Tucker. Dynamics of the stream-power river incision model: Implications for height limits of mountain ranges, landscape response time scales, and research needs. *J. Geophys. Res.*, 104:17661–17667, 1999.

[161] E. Wolanski and M. Eagle. Oceanography and fine sediment transport, fly river estuary and gulf of papua. In *Proceedings, 10th Australian Conference on Coastal & Ocean Engineering*, pages 453–457, Auckland, 1991.

[162] S. Wolfram. *A new kind of science*. Wolfram Media Inc., Champaign, Illinois, USA, 2002.

[163] L. D. Wright and J. M. Coleman. Variations in morphology of major river deltas as functions of ocean wave and river discharge regimes. *A.A.P.G. Bull.*, 57(2):370–398, 1973.

[164] L.D. Wright. River deltas. In Richard A. Jr. Davis, editor, *Coastal Sedimentary Environments*, pages 1–76. Springer-Verlag, New York, 1985.

[165] L.D. Wright, J. M. Coleman, and B.G. Thom. Process of channel development in a high tidal range environment: Camebridge gulf-ord river delta, western australia. *J. Geol.*, 81:15–41, 1973.

[166] L.D. Wright, J. M. Coleman, and B.G. Thom. Sediment transport and deposition in a macrotidal river channel: Ord river, western australia. In *Estuarine research*, volume 2, pages 309–321. Academic Press, 1975.

[167] S. Wright and G. Parker. Modeling downstream fining in sand-bed rivers, i/ii: Formulation. *J. Hydr. Res.*, 43(6):612–630, 2005.

Die VDM Verlagsservicegesellschaft sucht für wissenschaftliche Verlage abgeschlossene und herausragende

Dissertationen, Habilitationen, Diplomarbeiten, Master Theses, Magisterarbeiten usw.

für die kostenlose Publikation als Fachbuch.

Sie verfügen über eine Arbeit, die hohen inhaltlichen und formalen Ansprüchen genügt, und haben Interesse an einer honorarvergüteten Publikation?

Dann senden Sie bitte erste Informationen über sich und Ihre Arbeit per Email an *info@vdm-vsg.de*.

Sie erhalten kurzfristig unser Feedback!

VDM Verlagsservicegesellschaft mbH
Dudweiler Landstr. 99   Telefon +49 681 3720 174
D - 66123 Saarbrücken   Fax   +49 681 3720 1749

www.vdm-vsg.de

Die VDM Verlagsservicegesellschaft mbH vertritt

Printed by Books on Demand GmbH, Norderstedt / Germany